Animal Liberation

Also by Peter Singer

Democracy and Disobedience
Practical Ethics
Marx
The Expanding Circle
Animal Factories (with Jim Mason)
Hegel
Making Babies (with Deane Wells)
Should the Baby Live? (with Helga Kuhse)
Animal Liberation: A Graphic Guide (with Lori Gruen)

Animal Rights and Human Obligations (editor, with Tom Regan)
In Defense of Animals (editor)
Applied Ethics (editor)

Peter Singer

Animal Liberation

Second Edition

A New York Review Book
DISTRIBUTED BY RANDOM HOUSE, INC.

A **New York Review** Book

Distributed by Random House, Inc.
Published by The New York Review of Books
250 West 57th Street
New York, New York 10107

Cover design: M & Co.

Library of Congress Cataloging-in-Publication Data

Singer, Peter.
Animal liberation.

Includes bibliographical references.
1. Animal welfare. 2. Vegetarianism. I. Title
HV4708.S56 1990 179'.3 89-43160
ISBN 0-940322-00-5

First Printing, October 1975
Second Edition, January 1990

Printed in the U.S.A.

To Richard and Mary, and Ros and Stan,
and—especially—to Renata

This revised edition is also for all of you who have changed your lives in order to bring Animal Liberation closer. You have made it possible to believe that the power of ethical reasoning can prevail over the self-interest of our species.

Contents

Preface to the 1975 Edition

This book is about the tyranny of human over nonhuman animals. This tyranny has caused and today is still causing an amount of pain and suffering that can only be compared with that which resulted from the centuries of tyranny by white humans over black humans. The struggle against this tyranny is a struggle as important as any of the moral and social issues that have been fought over in recent years.

Most readers will take what they have just read to be a wild exaggeration. Five years ago I myself would have laughed at the statements I have now written in complete seriousness. Five years ago I did not know what I know today. If you read this book carefully, paying special attention to the second and third chapters, you will then know as much of what I know about the oppression of animals as it is possible to get into a book of reasonable length. Then you will be able to judge if my opening paragraph is a wild exaggeration or a sober estimate of a situation largely unknown to the general public. So I do not ask you to believe my opening paragraph now. All I ask is that you reserve your judgment until you have read the book.

Soon after I began work on this book my wife and I were invited to tea—we were living in England at the time—by a lady who had heard that I was planning to write about animals. She herself was very interested in animals, she said, and she had a friend who had already written a book about animals and would be *so* keen to meet us.

When we arrived our hostess's friend was already there, and she certainly was keen to talk about animals. "I do love animals," she began. "I have a dog and two cats, and do you know they get on together wonderfully well. Do you know Mrs. Scott? She runs a little hospital for sick pets . . . " and she was off. She paused while refreshments were served, took a ham sandwich, and then asked us what pets we had.

We told her we didn't own any pets. She looked a little surprised, and took a bite of her sandwich. Our hostess, who had now finished serving the sandwiches, joined us and took up the conversation: "But you *are* interested in animals, aren't you, Mr. Singer?"

We tried to explain that we were interested in the prevention of suffering and misery; that we were opposed to arbitrary discrimination; that we thought it wrong to inflict needless suffering on another being, even if that being were not a member of our own species; and that we believed animals were ruthlessly and cruelly exploited by humans, and we wanted this changed. Otherwise, we said, we were not especially "interested in" animals. Neither of us had ever been inordinately fond of dogs, cats, or horses in the way that many people are. We didn't "love" animals. We simply wanted them treated as the independent sentient beings that they are, and not as a means to human ends—as the pig whose flesh was now in our hostess's sandwiches had been treated.

This book is not about pets. It is not likely to be comfortable reading for those who think that love for animals involves no more than stroking a cat or feeding the birds in the garden. It is intended rather for people who are concerned about ending oppression and exploitation wherever they occur, and in seeing that the basic moral principle of equal consideration of interests is not arbitrarily restricted to members of our own species. The assumption that in order to be interested in such matters one must be an "animal-lover" is itself an indication of the absence of the slightest inkling that the moral standards that we apply among human beings might extend to other animals. No one, except a racist concerned to smear his opponents as "nigger-lovers," would suggest that in order to be concerned about equality for mistreated racial minorities you have to love those minorities, or regard them as cute and cuddly. So why make this assumption about people who work for improvements in the conditions of animals?

The portrayal of those who protest against cruelty to animals as sentimental, emotional "animal-lovers" has had the effect of excluding the entire issue of our treatment of nonhumans from serious political and moral discussion. It is easy to see why we do this. If we did give the issue serious consideration, if, for instance, we looked closely at the conditions in which animals live in the modern "factory farms" that produce our meat, we might be made uncomfortable about ham sandwiches, roast beef, fried chicken, and all those other items in our diet that we prefer not to think of as dead animals.

This book makes no sentimental appeals for sympathy toward "cute" animals. I am no more outraged by the slaughter of horses or dogs for meat than I am by the slaughter of pigs for this purpose. When the United States Defense Department finds that its use of beagles to test lethal gases has evoked a howl of protest and offers to use rats instead, I am not appeased.

This book is an attempt to think through, carefully and consistently, the question of how we ought to treat nonhuman animals. In the process it exposes the prejudices that lie behind our present attitudes and behavior. In the chapters that describe what these attitudes mean in practical terms—how animals suffer from the tyranny of human beings—there are passages that will arouse some emotions. These will, I hope, be emotions of anger and outrage, coupled with a determination to do something about the practices described. Nowhere in this book, however, do I appeal to the reader's emotions where they cannot be supported by reason. When there are unpleasant things to be described it would be dishonest to try to describe them in some neutral way that hid their real unpleasantness. You cannot write objectively about the experiments of the Nazi concentration camp "doctors" on those they considered "subhuman" without stirring emotions; and the same is true of a description of some of the experiments performed today on nonhumans in laboratories in America, Britain, and elsewhere. The ultimate justification for opposition to both these kinds of experiments, though, is not emotional. It is an appeal to basic moral principles which we all accept, and the application of these principles to the victims of both kinds of experiment is demanded by reason, not emotion.

The title of this book has a serious point behind it. A liberation movement is a demand for an end to prejudice and discrimination based on an arbitrary characteristic like race or sex. The classic instance is the Black Liberation movement. The immediate appeal of this movement, and its initial, if limited, success, made it a model for other oppressed groups. We soon became familiar with Gay Liberation and movements on behalf of American Indians and Spanish-speaking Americans. When a majority group—women—began their campaign some thought we had come to the end of the road. Discrimination on the basis of sex, it was said, was the last form of discrimination to be universally accepted and practiced without secrecy or pretense, even in those liberal circles that have long prided themselves on their freedom from prejudice against racial minorities.

We should always be wary of talking of "the last remaining form of discrimination." If we have learned anything from the liberation movements we should have learned how difficult it is to be aware of latent prejudices in our attitudes to particular groups until these prejudices are forcefully pointed out to us.

A liberation movement demands an expansion of our moral horizons. Practices that were previously regarded as natural and inevitable come to be seen as the result of an unjustifiable prejudice. Who can say with any confidence that none of his or her attitudes and practices can legitimately be questioned? If we wish to avoid being numbered among the oppressors, we must be prepared to rethink all our attitudes to other groups, including the most fundamental of them. We need to consider our attitudes from the point of view of those who suffer by them, and by the practices that follow from them. If we can make this unaccustomed mental switch we may discover a pattern in our attitudes and practices that operates so as consistently to benefit the same group—usually the group to which we ourselves belong—at the expense of another group. So we come to see that there is a case for a new liberation movement.

The aim of this book is to lead you to make this mental switch in your attitudes and practices toward a very large group of beings: members of species other than our own. I believe that our present attitudes to these beings are based on a long history of prejudice and arbitrary discrimination. I argue that there can be no reason—except the selfish desire to preserve the privileges of the exploiting group—for refusing to extend the basic principle of equality of consideration to members of other species. I ask you to recognize that your attitudes to members of other species are a form of prejudice no less objectionable than prejudice about a person's race or sex.

In comparison with other liberation movements, Animal Liberation has a lot of handicaps. First and most obvious is the fact that members of the exploited group cannot themselves make an organized protest against the treatment they receive (though they can and do protest to the best of their abilities individually). We have to speak up on behalf of those who cannot speak for themselves. You can appreciate how serious this handicap is by asking yourself how long blacks would have had to wait for equal rights if they had not been able to stand up for themselves and demand it. The less able a group is to stand up and organize against oppression, the more easily it is oppressed.

More significant still for the prospects of the Animal Liberation movement is the fact that almost all of the oppressing group are directly involved in, and see themselves as benefiting from, the oppression. There are few humans indeed who can view the oppression of animals with the detachment possessed, say, by Northern whites debating the institution of slavery in the Southern states of the Union. People who eat pieces of slaughtered nonhumans every day find it hard to believe that they are doing wrong; and they also find it hard to imagine what else they could eat. On this issue, anyone who eats meat is an interested party. They benefit—or at least they think they benefit—from the present disregard of the interests of nonhuman animals. This makes persuasion more difficult. How many Southern slaveholders were persuaded by the arguments used by the Northern abolitionists, and accepted by nearly all of us today? Some, but not many. I can and do ask you to put aside your interest in eating meat when considering the arguments of this book; but I know from my own experience that with the best will in the world this is not an easy thing to do. For behind the mere momentary desire to eat meat on a particular occasion lie many years of habitual meat-eating which have conditioned our attitudes to animals.

Habit. That is the final barrier that the Animal Liberation movement faces. Habits not only of diet but also of thought and language must be challenged and altered. Habits of thought lead us to brush aside descriptions of cruelty to animals as emotional, for "animal-lovers only"; or if not that, then anyway the problem is so trivial in comparison to the problems of human beings that no sensible person could give it time and attention. This too is a prejudice—for how can one know that a problem is trivial until one has taken the time to examine its extent? Although in order to allow a more thorough treatment this book deals with only two of the many areas in which humans cause other animals to suffer, I do not think anyone who reads it to the end will ever again think that the only problems that merit time and energy are problems concerning humans.

The habits of thought that lead us to disregard the interests of animals can be challenged, as they are challenged in the following pages. This challenge has to be expressed in a language, which in this case happens to be English. The English language, like other languages, reflects the prejudices of its users. So authors who wish to challenge these prejudices are in a well-known type of bind: either they use language that reinforces the very prejudices they wish to challenge, or else they fail to communicate with their audience. This book has already been forced along the former of these paths. We commonly use the word "animal" to mean "animals other than human beings." This usage sets humans apart from other animals, implying that we are not ourselves animals—an implication that everyone who has had elementary lessons in biology knows to be false.

In the popular mind the term "animal" lumps together beings as different as oysters and chimpanzees, while placing a gulf between chimpanzees and humans, although our relationship to those apes is much closer than the oyster's. Since there exists no other short term for the nonhuman animals, I have, in the title of this book and elsewhere in these pages, had to use "animal" as if it did not include the human animal. This is a regrettable lapse from the standards of revolutionary purity but it seems necessary for effective communication. Occasionally, however, to remind you that this is a matter of convenience only, I shall use longer, more accurate modes of referring to what was once called "the brute creation." In other cases, too, I have tried to avoid language which tends to degrade animals or disguise the nature of the food we eat.

The basic principles of Animal Liberation are very simple. I have tried to write a book that is clear and easy to understand, requiring no expertise of any kind. It is necessary, however, to begin with a discussion of the principles that underlie what I have to say. While there should be nothing here that is difficult, readers unused to this kind of discussion might find the first chapter rather abstract. Don't be put off. In the next chapters we get down to the little-known details of how our species oppresses others under our control. There is nothing abstract about this oppression, or about the chapters that describe it.

If the recommendations made in the following chapters are accepted, millions of animals will be spared considerable pain. Moreover, millions of humans will benefit too. As I write, people are starving to death in many parts of the world; and many more are in imminent danger of starvation. The United States government has said that because of poor harvests and diminished stocks of grain it can provide only limited—and inadequate—assistance; but as Chapter 4 of this book makes clear, the heavy emphasis in affluent nations on rearing animals for food wastes several times as much food as it produces. By ceasing to rear and kill animals for food, we can make so much extra food available for humans that, properly distributed, it would eliminate starvation and malnutrition from this planet. Animal Liberation is Human Liberation too.

Preface to the New Edition

To reread the original preface of this book is to return to a world half forgotten. People concerned about animals don't offer me ham sandwiches anymore. In Animal Liberation groups, the activists are now all vegetarian; but even in the more conservative animal welfare movement, there is some awareness of the issue of eating animals. Those who do so are apologetic about it and ready to provide alternatives when preparing meals for others. A new consciousness exists about the need to extend sympathies for dogs and cats to pigs, chickens, and even laboratory rats.

I am not sure how much credit *Animal Liberation* can take for this change. Popular magazine writers have given it the tag line "the bible of the animal liberation movement." It is a line that I cannot help finding flattering, but it makes me uncomfortable at the same time. I don't believe in bibles: no book has a monopoly on truth. In any case, no book can achieve anything unless it strikes a chord in its readers. The liberation movements of the Sixties had made Animal Liberation an obvious next step: this book drew the arguments together and gave them a coherent shape. The rest was done by some very fine, ethically concerned, hard-working people—first a few individuals, then hundreds, gradually growing into thousands and now perhaps millions—who make up the Animal Liberation movement. I have dedicated this revised edition to them, because without them the first edition would have suffered the fate of Henry Salt's book *Animals' Rights*, published in 1892 and left to gather dust on the shelves of the British Museum library until, eighty years later, a new generation formulated the arguments afresh, stumbled across a few obscure references, and discovered that it had all been said before, but to no avail.

This time it will not be in vain. The movement has grown too big for that. Important gains for animals have already been achieved. Far greater ones lie ahead. Animal Liberation is now a worldwide movement, and it will be on the agenda for a long time to come.

People often ask me if I am pleased about the way the movement has grown. The way they ask the question makes it plain that they are expecting me to say that I never dreamed that the book would have such an impact. But they are wrong. In my dreams, at least, everyone who read the book was going to say, "Yes, of course . . . ," and would immediately become vegetarian, and start protesting against what we do to animals, so that more people would hear about the message of Animal Liberation, and at least the most extreme and needless forms of animal suffering would soon be stopped by an irresistible tide of public protest.

Admittedly, such dreams were balanced by my awareness of the obstacles: the conservatism of most of us when it comes to what we put into our stomachs; the financial interests that would fight to the last million to defend their right to exploit animals for maximum profit; and the solid weight of history and tradition bolstering the attitudes that justify this exploitation. So I am pleased to have met, and received letters from, many, many people who did read the book and say, "Yes, of course . . . ," and did stop eating animals, and became active in the Animal Liberation movement. I am even more pleased, of course, that after many years' hard struggle by many people, the Animal Liberation movement is a political and social reality. But, even so, that is not enough; not nearly enough. As this edition shows only too clearly, the movement has as yet made little impact on the central forms of animal exploitation.

Animal Liberation was first published in 1975 and has remained in print, virtually unaltered, since then. Three aspects are now ripe for revision. First, when the book appeared there was no Animal Liberation movement. The term itself was unknown and there were no large organizations—and hardly any small ones—working to implement radical changes in our attitudes and practices toward animals. Fifteen years later it is decidedly odd to have a book entitled *Animal Liberation* that takes no notice of the existence of the modern Animal Liberation movement, and for that reason makes no comment on the course the movement has taken.

Secondly, the rise of the modern Animal Liberation movement has been paralleled by an amazing increase in the amount that has been written on this topic—much of it commenting on the position taken in the first edition of this book. I have also spent long evenings discussing both philosophical issues and practical conclusions with friends and fellow workers within the Animal Liberation movement. Some response to all this discussion seemed necessary, if only to indicate the extent to which I have or have not altered my views.

Finally, the second and third chapters of this book describe what our present attitudes mean for animals in two major fields of animal use: experimentation and farming. As soon as I began hearing people saying things like "Of course, things have improved a lot since that was written . . ." I knew that it was necessary to document what is happening now in laboratories and farms and to present readers with descriptions that cannot be palmed off as belonging to some distant dark age.

The new descriptions make up most of the differences between this and the previous edition. I have, however, resisted suggestions that I should add similar accounts of other kinds of animal abuse. The aim of the factual material is not to serve as a comprehensive report on how we treat animals; it is rather, as I indicate at the end of the first chapter, to display in sharp, clear, and concrete form the implications of the more abstract philosophical conception of speciesism presented in the first chapter. The omission of discussions of hunting and trapping, of the fur industry, of the abuse of companion animals, of rodeos, zoos, and circuses means not that these matters are less important, but only that the two central cases of experimentation and food production are sufficient to serve my purpose.

I decided against trying to respond to all the points raised by philosophers about the ethical arguments of this book. To do so would have changed the nature of the book itself, turning it into a work of academic philosophy, of interest to my professional colleagues but tedious for the general reader. Instead I have indicated, at appropriate places in the text, some other writings where my replies to certain objections can be found. I have also rewritten one passage, in the final chapter, in which I have changed my mind about a philosophical point that has only a peripheral relationship to the ethical foundation on which the argument of this book rests. As for these foundations themselves, I have lectured on them, given talks to conferences and philosophy department seminars, and discussed them at length, both verbally and in print; but I have come across no insurmountable objections, nothing that has led me to think that the simple ethical arguments on which the book is based are anything but sound. I have been encouraged to find that many of my most respected philosophical colleagues agree. Hence these arguments are retained here, unchanged.

This leaves the first of the three aspects of the book that are in need of updating, which I mentioned above: an account of the Animal Liberation movement and the course it has taken.

Both in the accounts of laboratory experimentation and factory farming and in the final chapter of this revised edition, I refer to some of the major campaigns and achievements of the Animal Liberation movement. I have not attempted to describe the campaigns in detail, because some of the leading activists did so themselves, in a book called *In Defense of Animals*, which I edited not long ago. But one issue of importance to the movement needs to be raised in a prominent place in this book, and I do it here. That issue is violence.

Activists have practiced a variety of means for advancing toward the goal of Animal Liberation. Some seek to educate the public by distributing leaflets and writing letters to newspapers. Others lobby government officials and their elected representatives in Parliament or Congress. Activist organizations hold demonstrations and protest outside places where animals are being made to suffer to serve trivial human objectives. But many become impatient with the slow progress made by such means and want to take more direct action to stop the suffering now.

No one who understands just what animals are enduring can be critical of such impatience. In the face of a continuing atrocity it is scarcely enough to sit back and write letters. The need is to help the animals now. But how? The usual legitimate channels for political protest are slow and uncertain. Should one break in and free the animals? That is illegal, but the obligation to obey the law is not absolute. It was justifiably broken by those who helped runaway slaves in the American South, to mention only one possible parallel. A more serious problem is that the literal liberation of animals from laboratories and factory farms can only be a token gesture, for the researchers will simply order another batch of animals, and who can find homes for a thousand factory farm pigs or 100,000 hens? Raids by Animal Liberation Front groups in several countries have been more effective when they have obtained evidence of animal abuse that could not otherwise have come to light. In the case of the raid on Dr. Thomas Gennarelli's laboratory at the University of Pennsylvania, for example, stolen videotapes provided the evidence that finally convinced even the secretary for health and human services that the experiments must stop. It is hard to imagine that this result could have been achieved in any other way, and I have nothing but praise for the courageous, caring, and thoughtful people who planned and carried out that particular action.

But other illegal activities are very different. In 1982 a group calling itself the "Animal Rights Militia" sent letter-bombs to Margaret Thatcher; and in 1988 Fran Trutt, an animal activist, was arrested while planting a bomb outside the offices of the U.S. Surgical Corporation, a company that had been using live dogs to demonstrate its surgical stapling devices. Neither of these actions was in any way representative of the Animal Liberation movement. The Animal Rights Militia had never been heard of before, and was immediately condemned by all the organizations of the British Animal Liberation movement. Trutt was working alone, and her actions received a quick denunciation from the American movement. (The evidence also suggested entrapment, for she was driven to the corporation's offices by a paid undercover informant hired by U.S. Surgical's security consultant.) Yet such actions can be seen as the extreme end of a spectrum of threats and harassment of experimenters, furriers, and other animal exploiters, and so it is important that those within the Animal Liberation movement make their position toward such actions plain.

It would be a tragic mistake if even a small section of the Animal Liberation movement were to attempt to achieve its objectives by hurting people. Some believe that people who make animals suffer deserve to have suffering inflicted upon them. I don't believe in vengeance; but even if I did, it would be a damaging distraction from our task of stopping the suffering. To do that, we must change the minds of reasonable people in our society. We may be convinced that a person who is abusing animals is entirely callous and insensitive; but we lower ourselves to that level if we physically harm or threaten physical harm to that person. Violence can only breed more violence—a cliché, but one that can be seen to be tragically true in half a dozen conflicts around the world. The strength of the case for Animal Liberation is its ethical commitment; we occupy the high moral ground and to abandon it is to play into the hands of those who oppose us.

The alternative to the path of increasing violence is to follow the lead of the two greatest—and not coincidentally, most successful—leaders of liberation movements in modern times: Gandhi and Martin Luther King. With immense courage and resolution, they stuck to the principle of nonviolence despite the provocations, and often violent attacks, of their opponents. In the end they succeeded because the justice of their cause could not be denied, and their behavior touched the consciences even of those who had opposed them. The wrongs we inflict on other species are equally undeniable once they are seen plainly; and it is in the rightness of our cause, and not the fear of our bombs, that our prospects of victory lie.

Acknowledgments

It is normal practice to thank those who have assisted in the writing of a book; but in the present instance my debts are of a special kind, which can only be indicated by a brief narrative.

In the fall of 1970 I was a graduate student at the University of Oxford. Although I had specialized in moral and social philosophy, it had not occurred to me—any more than it occurs to most people—that our relations with animals raised a serious moral issue. I knew, of course, that some animals were cruelly treated, but I assumed that these were incidental abuses and not an indication of anything fundamentally wrong.

My complacency was disturbed when I met Richard Keshen, a fellow student at Oxford and a vegetarian. Over lunch I asked him why he did not eat meat, and he began to tell me about the conditions in which the animal whose body I was eating had lived. Through Richard and his wife Mary, my wife and I became friendly with Roslind and Stanley Godlovitch, also vegetarians studying philosophy at Oxford. In long conversations with these four—and particularly with Roslind Godlovitch, who had worked out her ethical position in considerable detail—I became convinced that by eating animals I was participating in a systematic form of oppression of other species by my own species. The central ideas of this book derive from these conversations.

Reaching a theoretical conclusion is one thing; putting it into practice is another. Without the support and encouragement of Renata, my wife, who had become equally convinced that our friends were right, I might still be eating meat, though with a guilty conscience.

The idea of writing a book arose from the enthusiastic response to my review of *Animals, Men and Morals*, edited by Stanley and Roslind Godlovitch and John Harris, which appeared in *The New York Review of Books* (April 5, 1973). I am grateful to the editors of *The New York Review* for publishing this unsolicited discussion of a book on an unfashionable topic. The review would, however, never have turned into a book without the encouragement and assistance of the following:

Eleanor Seiling of United Action for Animals, New York, made her organization's unique collection of documents on experimental uses of animals available to me; and Alois Acowitz's summaries of the experimenters' reports enabled me to find what I wanted in a fraction of the time that would otherwise have been needed.

Richard Ryder generously lent me material he had gathered for his own book, *Victims of Science*.

Joanne Bower, of the Farm and Food Society, London, provided me with information on the conditions of farm animals in Britain.

Kathleen Jannaway, of the Vegan Society of the United Kingdom, helped me to locate reports on the nutritional adequacy of plant foods.

John Norton, of the Animal Rescue League of Boston, and Martha Coe, of Argus Archives, New York, provided materials on the transport and slaughter of animals in the United States.

The Scottish Society for the Prevention of Vivisection was of assistance in obtaining photographs of experiments on animals.

Dudley Giehl, of Animal Liberation, Inc., New York, allowed me to use material he had collected on intensive farming and vegetarianism.

Alice Herrington and Joyce Lambert, of Friends of Animals, New York, assisted in many ways, and Jim Mason, of the same organization, arranged visits to intensive farms.

An invitation to take up a visiting position in the Department of Philosophy at New York University for the academic year 1973–1974 provided me with a congenial atmosphere and an ideal location for research and writing, and my colleagues and students gave me valuable comments and criticism. I also had the opportunity to subject my views on animals to the critical scrutiny of students and faculty of the philosophy departments at the following universities: Brown University, Fordham University, Long Island University, North Carolina State University at

Raleigh, Rutgers University, State University of New York at Brockport, State University of New York at Stony Brook, Tufts University, University of California at Berkeley, University of Miami, and Williams College; and at the Yale Law School and a meeting of the Society for Philosophy and Public Affairs, New York. Chapters 1 and 6 of this book benefited considerably from the discussions following my talks.

Finally I must thank the editors and publisher of *The New York Review of Books* for their support for the book, and especially Robert Silvers, whose thoughtful editorial advice has considerably improved the original manuscript. It is left only to add that responsibility for any remaining imperfections is mine alone.

P.S.
February 1975

Acknowledgments to the Revised Edition

So many people, from all over the world, have helped in preparing this revised edition that I am bound to leave some out, for which I can only apologize. Sometimes the assistance consisted of reading drafts, sometimes it was sending me material so that I could keep abreast of developments in many different countries. Here is a list, in no particular order: Don Barnes and Melinda Moreland of the (U.S.) National Anti-Vivisection Society, Alex Hershaft of Farm Animal Reform Movement, MacDonald White and Ann St. Laurent of United Action for Animals, Joyce D'Silva and Carol Long of Compassion in World Farming, Clare Druce and Violet Spalding of Chickens' Lib, Henry Spira of Animal Rights International, Brad Miller of the Humane Farming Association, Kim Stallwood and Carla Bennett of People for the Ethical Treatment of Animals, Peter Hamilton of Lifeforce, Maria Comninou of the Ann Arbor Association for Responsible Animal Treatment, George Cave of Trans-Species Unlimited, Paola Cavalieri of *Etica & Animali* in Milan, Birgitta Carlsson of the Swedish Society Against Painful Experiments on Animals, Detlef Fölsch of the Institute of Animal Sciences, Swiss Federal Institute of Technology, Charles Magel, John Robbins, Richard Ryder, Clive Hollands, and Jim Mason.

Special thanks are due to Lori Gruen, who worked as a kind of United States coordinator, gathering new material for, and helping me to update, the chapters on experimentation and factory farming. She also made many valuable suggestions on a draft of the entire book. Lori in turn wishes to thank, in addition to those listed above, the following who supplied information to her: Diane Halverson of the Animal Welfare Institute; Avi Magidoff, Jeff Diner, and Martin Stephens, whose work on aspects of animal experimentation in the United States proved an invaluable resource; and Ken Knowles and Dave Macauley.

My revisions to the chapter on factory farming were further aided by work done for a submission on factory farming, prepared with enormous industry and impeccable scholarship by Suzanne Pope and Geoff Russell for the Australian and New Zealand Federation of Animal Societies. My comments on fish and fishing benefited from another fine submission, prepared for Animal Liberation (Victoria) by Patty Mark.

Finally, I again owe a great deal to *The New York Review of Books:* to Robert Silvers for his support for the idea of a new edition and for the application of his fine critical skills in the process of editing it; to Rea Hederman, who directed the many facets of publication; and to Neil Gordon, who oversaw the typesetting with wonderful care and attention to detail.

P.S.
November 1989

Chapter 1

All Animals Are Equal . . .

or why the ethical principle on which human equality rests requires us to extend equal consideration to animals too

"Animal Liberation" may sound more like a parody of other liberation movements than a serious objective. The idea of "The Rights of Animals" actually was once used to parody the case for women's rights. When Mary Wollstonecraft, a forerunner of today's feminists, published her *Vindication of the Rights of Woman* in 1792, her views were widely regarded as absurd, and before long an anonymous publication appeared entitled *A Vindication of the Rights of Brutes*. The author of this satirical work (now known to have been Thomas Taylor, a distinguished Cambridge philosopher) tried to refute Mary Wollstonecraft's arguments by showing that they could be carried one stage further. If the argument for equality was sound when applied to women, why should it not be applied to dogs, cats, and horses? The reasoning seemed to hold for these "brutes" too; yet to hold that brutes had rights was manifestly absurd. Therefore the reasoning by which this conclusion had been reached must be unsound, and if unsound when applied to brutes, it must also be unsound when applied to women, since the very same arguments had been used in each case.

In order to explain the basis of the case for the equality of animals, it will be helpful to start with an examination of the case for the equality of women. Let us assume that we wish to defend the case for women's rights against the attack by Thomas Taylor. How should we reply?

One way in which we might reply is by saying that the case for equality between men and women cannot validly be extended to nonhuman animals. Women have a right to vote, for instance, because they are just as capable of making rational decisions about the future as men are; dogs, on the other hand, are incapable of understanding the significance of voting, so they cannot have the right to vote. There are many other obvious ways in which men and women resemble each other closely, while humans and animals differ greatly. So, it might be said, men and women are similar beings and should have similar rights, while humans and nonhumans are different and should not have equal rights.

The reasoning behind this reply to Taylor's analogy is correct up to a point, but it does not go far enough. There are obviously important differences between humans and other animals, and these differences must give rise to some differences in the rights that each have. Recognizing this evident fact, however, is no barrier to the case for extending the basic principle of equality to nonhuman animals. The differences that exist between men and women are equally undeniable, and the supporters of Women's Liberation are aware that these differences may give rise to different rights. Many feminists hold that women have the right to an abortion on request. It does not follow that since these same feminists are campaigning for equality between men and women they must support the right of men to have abortions too. Since a man cannot have an abortion, it is meaningless to talk of his right to have one. Since dogs can't vote, it is meaningless to talk of their right to vote. There is no reason why either Women's Liberation or Animal Liberation should get involved in such nonsense. The extension of the basic principle of equality from one group to another does not imply that we must treat both groups in exactly the same way, or grant exactly the same rights to both groups. Whether we should do so will depend on the nature of the members of the two groups. The basic principle of equality does not require equal or identical *treatment*; it requires equal consideration. Equal consideration for different beings may lead to different treatment and different rights.

So there is a different way of replying to Taylor's attempt to parody the case for women's rights, a way that does not deny the obvious differences between human beings and nonhumans but goes more deeply into the question of equality and concludes by finding nothing absurd in the idea that the basic principle of equality applies to so-called brutes. At this point such a conclusion may appear odd; but if we examine more deeply the basis on which our opposition to discrimination on grounds of race or sex ultimately rests, we will see that we would be on shaky ground if we were to demand equality for blacks, women, and other groups of oppressed humans while denying equal consideration to nonhumans. To make this clear we need to see, first, exactly why racism and sexism are wrong. When we say that all human beings, whatever their race, creed, or sex, are equal, what is it that we are asserting? Those who wish to defend hierarchical, inegalitarian societies have often pointed out that by whatever test we choose it simply is not true that all humans are equal. Like it or not we must face the fact that humans come in different shapes and sizes; they come with different moral capacities, different intellectual abilities, different amounts of benevolent feeling and sensitivity to the needs of others, different abilities to communicate effectively, and different capacities to experience pleasure and pain. In short, if the demand for equality were based on the actual equality of all human beings, we would have to stop demanding equality.

Still, one might cling to the view that the demand for equality among human beings is based on the actual equality of the different races and sexes. Although, it may be said, humans differ as individuals, there are no differences between the races and sexes as such. From the mere fact that a person is black or a woman we cannot infer anything about that person's intellectual or moral capacities. This, it may be said, is why racism and sexism are wrong. The white racist claims that whites are superior to blacks, but this is false; although there are differences among individuals, some blacks are superior to some whites in all of the capacities and abilities that could conceivably be relevant. The opponent of sexism would say the same: a person's sex is no guide to his or her abilities, and this is why it is unjustifiable to discriminate on the basis of sex.

The existence of individual variations that cut across the lines of race or sex, however, provides us with no defense at all against a more sophisticated opponent of equality, one who proposes that, say, the interests of all those with IQ scores below 100 be given less consideration than the interests of those with ratings over 100. Perhaps those scoring below the mark would, in this society, be made the slaves of those scoring higher. Would a hierarchical society of this sort really be so much better than one based on race or sex? I think not. But if we tie the moral principle of equality to the factual equality of the different races or sexes, taken as a whole, our opposition to racism and sexism does not provide us with any basis for objecting to this kind of inegalitarianism.

There is a second important reason why we ought not to base our opposition to racism and sexism on any kind of factual equality, even the limited kind that asserts that variations in capacities and abilities are spread evenly among the different races and between the sexes: we can have no absolute guarantee that these capacities and abilities really are distributed evenly, without regard to race or sex, among human beings. So far as actual abilities are concerned there do seem to be certain measurable differences both among races and between sexes. These differences do not, of course, appear in every case, but only when averages are taken. More important still, we do not yet know how many of these differences are really due to the different genetic endowments of the different races and sexes, and how many are due to poor schools, poor housing, and other factors that are the result of past and continuing discrimination. Perhaps all of the important differences will eventually prove to be environmental rather than genetic. Anyone opposed to racism and sexism will certainly hope that this will be so, for it will make the task of ending discrimination a lot easier; nevertheless, it would be dangerous to rest the case against racism and sexism on the belief that all significant differences are environmental in origin. The opponent of, say, racism who takes this line will be unable to avoid conceding that if differences in ability did after all prove to have some genetic connection with race, racism would in some way be defensible.

Fortunately there is no need to pin the case for equality to one particular outcome of a scientific investigation. The appropriate response to those who claim to have found evidence of genetically based differences in ability among the races or between the sexes is not to stick to the belief that the genetic explanation must be wrong, whatever evidence to the contrary may turn up; instead we should make it quite clear that the claim to equality does not depend on intelligence, moral capacity, physical strength, or similar matters of fact. Equality is a moral idea, not an assertion of fact. There is no logically compelling reason for assuming that a factual difference in ability between two people justifies any difference in the amount of consideration we give to their needs and interests. *The principle of the equality of human beings is not a description of an alleged actual equality among humans: it is a prescription of how we should treat human beings.*

Jeremy Bentham, the founder of the reforming utilitarian school of moral philosophy, incorporated the essential basis of moral equality into his system of ethics by means of the formula: "Each to count for one and none for more than one." In other words, the interests of every being affected by an action are to be taken into account and given the same weight as the like interests of any other being. A later utilitarian, Henry Sidgwick, put the point in this way: "The good of any one individual is of no more importance, from the point of view (if I may say so) of the Universe, than the good of any other." More recently the leading figures in contemporary moral philosophy have shown a great deal of agreement in specifying as a fundamental presupposition of their moral theories some similar requirement that works to give everyone's interests equal consideration—although these writers generally cannot agree on how this requirement is best formulated.[1]

It is an implication of this principle of equality that our concern for others and our readiness to consider their interests ought not to depend on what they are like or on what abilities they may possess. Precisely what our concern or consideration requires us to do may vary according to the characteristics of those affected by what we do: concern for the well-being of children growing up in America would require that we teach them to read; concern for the well-being of pigs may require no more than that we leave them with other pigs in a place where there is adequate food and room to run freely. But the basic element—the taking into account of the interests of the being, whatever those interests may be—must, according to the principle of equality, be extended to all beings, black or white, masculine or feminine, human or nonhuman.

Thomas Jefferson, who was responsible for writing the principle of the equality of men into the American Declaration of Independence, saw this point. It led him to oppose slavery even though he was unable to free himself fully from his slaveholding background. He wrote in a letter to the author of a book that emphasized the notable intellectual achievements of Negroes in order to refute the then common view that they had limited intellectual capacities:

> Be assured that no person living wishes more sincerely than I do, to see a complete refutation of the doubts I myself have entertained and expressed on the grade of understanding allotted to them by nature, and to find that they are on a par with ourselves . . . but whatever be their degree of talent it is no measure of their rights. Because Sir Isaac Newton was superior to others in understanding, he was not therefore lord of the property or persons of others.[2]

Similarly, when in the 1850s the call for women's rights was raised in the United States, a remarkable black feminist named Sojourner Truth made the same point in more robust terms at a feminist convention:

> They talk about this thing in the head; what do they call it? ["Intellect," whispered someone nearby.] That's it. What's that got to do with women's rights or Negroes' rights? If my cup won't hold but a pint and yours holds a quart, wouldn't you be mean not to let me have my little half-measure full?[3]

It is on this basis that the case against racism and the case against sexism must both ultimately rest; and it is in accordance with this principle that the attitude that we may call "speciesism," by analogy with racism, must also be condemned. Speciesism—the word is not an attractive one, but I can think of no better term—is a prejudice or attitude of bias in favor of the interests of members of one's own species and against those of members of other species. It should be obvious that the fundamental objections to racism and sexism made by Thomas Jefferson and Sojourner Truth apply equally to speciesism. If possessing a higher degree of intelligence does not entitle one human to use another for his or her own ends, how can it entitle humans to exploit nonhumans for the same purpose?[4]

Many philosophers and other writers have proposed the principle of equal consideration of interests, in some form or other, as a basic moral principle; but not many of them have recognized that this principle applies to members of other species as well as to our own. Jeremy Bentham was one of the few who did realize this. In a forward-looking passage written at a time when black slaves had been freed by the French but in the British dominions were still being treated in the way we now treat animals, Bentham wrote:

> The day *may* come when the rest of the animal creation may acquire those rights which never could have been withholden from them but by the hand of tyranny. The French have already discovered that the blackness of the skin is no reason why a human being should be abandoned without redress to the caprice of a tormentor. It may one day come to be recognized that the number of the legs, the villosity of the skin, or the termination of the *os sacrum* are reasons equally insufficient for abandoning a sensitive being to the same fate. What else is it that should trace the insuperable line? Is it the faculty of reason, or perhaps the faculty of discourse? But a full-grown horse or dog is beyond comparison a more rational, as well as a more conversable animal, than an infant of a day or a week or even a month, old. But suppose they were otherwise, what would it avail? The question is not, Can they *reason*? nor Can they *talk*? but, Can they *suffer*?[5]

In this passage Bentham points to the capacity for suffering as the vital characteristic that gives a being the right to equal consideration. The capacity for suffering—or more strictly, for suffering and/or enjoyment or happiness—is not just another characteristic like the capacity for language or higher mathematics. Bentham is not saying that those who try to mark "the insuperable line" that determines whether the interests of a being should be considered happen to have chosen the wrong characteristic. By saying that we must consider the interests of all beings with the capacity for suffering or enjoyment Bentham does not arbitrarily exclude from consideration any interests at all—as those who draw the line with reference to the possession of reason or language do. The capacity for suffering and enjoyment is *a prerequisite for having interests at all,* a condition that must be satisfied before we can speak of interests in a meaningful way. It would be nonsense to say that it was not in the interests of a stone to be kicked along the road by a schoolboy. A stone does not have interests because it cannot suffer. Nothing that we can do to it could possibly make any difference to its welfare. The capacity for suffering and enjoyment is, however, not only necessary, but also sufficient for us to say that a being has interests—at an absolute minimum, an interest in not suffering. A mouse, for example, does have an interest in not being kicked along the road, because it will suffer if it is.

Although Bentham speaks of "rights" in the passage I have quoted, the argument is really about equality rather than about rights. Indeed, in a different passage, Bentham famously described "natural rights" as "nonsense" and "natural and imprescriptable rights" as "nonsense upon stilts." He talked of moral rights as a shorthand way of referring to protections that people and animals morally ought to have; but the real weight of the moral argument does not rest on the assertion of the existence of the right, for this in turn has to be justified on the basis of the possibilities for suffering and happiness. In this way we can argue for equality for animals without getting embroiled in philosophical controversies about the ultimate nature of rights.

In misguided attempts to refute the arguments of this book, some philosophers have gone to much trouble developing arguments to show that animals do not have rights.[6] They have claimed that to have rights a being must be autonomous, or must be a member of a community, or must have the ability to respect the rights of others, or must possess a sense of justice. These claims are irrelevant to the case for Animal Liberation. The language of rights is a convenient political shorthand. It is even more valuable in the era of thirty-second TV news clips than it was in Bentham's day; but in the argument for a radical change in our attitude to animals, it is in no way necessary.

If a being suffers there can be no moral justification for refusing to take that suffering into consideration. No matter what the nature of the being, the principle of equality requires that its suffering be counted equally with the like suffering—insofar as rough comparisons can be made—of any other being. If a being is not capable of suffering, or of experiencing enjoyment or happiness, there is nothing to be taken into account. So the limit of sentience (using the term as a convenient if not strictly accurate shorthand for the capacity to suffer and/or experience enjoyment) is the only defensible boundary of concern for the interests of others. To mark this boundary by some other characteristic like intelligence or rationality would be to mark it in an arbitrary manner. Why not choose some other characteristic, like skin color?

Racists violate the principle of equality by giving greater weight to the interests of members of their own race when there is a clash between their interests and the interests of those of another race. Sexists violate the principle of equality by favoring the interests of their own sex. Similarly, speciesists allow the interests of their own species to override the greater interests of members of other species. The pattern is identical in each case.

Most human beings are speciesists. The following chapters show that ordinary human beings—not a few exceptionally cruel or heartless humans, but the overwhelming majority of humans—take an active part in, acquiesce in, and allow their taxes to pay for practices that require the sacrifice of the most important interests of members of other species in order to promote the most trivial interests of our own species.

There is, however, one general defense of the practices to be described in the next two chapters that needs to be disposed of before we discuss the practices themselves. It is a defense which, if true, would allow us to do anything at all to nonhumans for the slightest reason, or for no reason at all, without incurring any justifiable reproach. This defense claims that we are never guilty of neglecting the interests of other animals for one breathtakingly simple reason: they have no interests. Nonhuman animals have no interests, according to this view, because they are not capable of suffering. By this is not meant merely that they are not capable of suffering in all the ways that human beings are—for instance, that a calf is not capable of suffering from the knowledge that it will be killed in six months time. That modest claim is, no doubt, true; but it does not clear humans of the charge of speciesism, since it allows that animals may suffer in other ways—for instance, by being given electric shocks, or being kept in small, cramped cages. The defense I am about to discuss is the much more sweeping, although correspondingly less plausible, claim that animals are incapable of suffering in any way at all; that they are, in fact, unconscious automata, possessing neither thoughts nor feelings nor a mental life of any kind.

Although, as we shall see in a later chapter, the view that animals are automata was proposed by the seventeenth-century French philosopher René Descartes, to most people, then and now, it is obvious that if, for example, we stick a sharp knife into the stomach of an unanesthetized dog, the dog will feel pain. That this is so is assumed by the laws in most civilized countries that prohibit wanton cruelty to animals. Readers whose common sense tells them that animals do suffer may prefer to skip the remainder of this section, moving straight on to page 15, since the pages in between do nothing but refute a position that they do not hold. Implausible as it is, though, for the sake of completeness this skeptical position must be discussed.

Do animals other than humans feel pain? How do we know? Well, how do we know if anyone, human or nonhuman, feels pain? We know that we ourselves can feel pain. We know this from the direct experience of pain that we have when, for instance, somebody presses a lighted cigarette against the back of our hand. But how do we know that anyone else feels pain? We cannot directly experience anyone else's pain, whether that "anyone" is our best friend or a stray dog. Pain is a state of consciousness, a "mental event," and as such it can never be observed. Behavior like writhing, screaming, or drawing one's hand away from the lighted cigarette is not pain itself; nor are the recordings a neurologist might make of activity within the brain observations of pain itself. Pain is something that we feel, and we can only infer that others are feeling it from various external indications.

In theory, we *could* always be mistaken when we assume that other human beings feel pain. It is conceivable that one of our close friends is really a cleverly constructed robot, controlled by a brilliant scientist so as to give all the signs of feeling pain, but really no more sensitive than any other machine. We can never know, with absolute certainty, that this is not the case. But while this might present a puzzle for philosophers, none of us has the slightest real doubt that our close friends feel pain just as we do. This is an inference, but a perfectly reasonable one, based on observations of their behavior in situations in which we would feel pain, and on the fact that we have every reason to assume that our friends are beings like us, with nervous systems like ours that can be assumed to function as ours do and to produce similar feelings in similar circumstances.

If it is justifiable to assume that other human beings feel pain as we do, is there any reason why a similar inference should be unjustifiable in the case of other animals?

Nearly all the external signs that lead us to infer pain in other humans can be seen in other species, especially the species most closely related to us—the species of mammals and birds. The behavioral signs include writhing, facial contortions, moaning, yelping or other forms of calling, attempts to avoid the source of pain, appearance of fear at the prospect of its repetition, and so on. In addition, we know that these animals have nervous systems very like ours, which respond physiologically as ours do when the animal is in circumstances in which we would feel pain: an initial rise of blood pressure, dilated pupils, perspiration, an increased pulse rate, and, if the stimulus continues, a fall in blood pressure. Although human beings have a more developed cerebral cortex than other animals, this part of the brain is concerned with thinking functions rather than with basic impulses, emotions, and feelings. These impulses, emotions, and feelings are located in the diencephalon, which is well developed in many other species of animals, especially mammals and birds.[7]

We also know that the nervous systems of other animals were not artificially constructed—as a robot might be artificially constructed—to mimic the pain behavior of humans. The nervous systems of animals evolved as our own did, and in fact the evolutionary history of human beings and other animals, especially mammals, did not diverge until the central features of our nervous systems were already in existence. A capacity to feel pain obviously enhances a species' prospects of survival, since it causes members of the species to avoid sources of injury. It is surely unreasonable to suppose that nervous systems that are virtually identical physiologically, have a common origin and a common evolutionary function, and result in similar forms of behavior in similar circumstances should actually operate in an entirely different manner on the level of subjective feelings.

It has long been accepted as sound policy in science to search for the simplest possible explanation of whatever it is we are trying to explain. Occasionally it has been claimed that it is for this reason "unscientific" to explain the behavior of animals by theories that refer to the animal's conscious feelings, desires, and so on—the idea being that if the behavior in question can be explained without invoking consciousness or feelings, that will be the simpler theory. Yet we can now see that such explanations, when assessed with respect to the actual behavior of both human and nonhuman animals, are actually far more complex than rival explanations. For we know from our own experience that explanations of our own behavior that did not refer to consciousness and the feeling of pain would be incomplete; and it is simpler to assume that the similar behavior of animals with similar nervous systems is to be explained in the same way than to try to invent some other explanation for the behavior of nonhuman animals as well as an explanation for the divergence between humans and nonhumans in this respect.

The overwhelming majority of scientists who have addressed themselves to this question agree. Lord Brain, one of the most eminent neurologists of our time, has said:

> I personally can see no reason for conceding mind to my fellow men and denying it to animals.... I at least cannot doubt that the interests and activities of animals are correlated with awareness and feeling in the same way as my own, and which may be, for aught I know, just as vivid.[8]

The author of a book on pain writes:

> Every particle of factual evidence supports the contention that the higher mammalian vertebrates experience pain sensations at least as acute as our own. To say that they feel less because they are lower animals is an absurdity; it can easily be shown that many of their senses are far more acute than ours—visual acuity in certain birds, hearing in most wild animals, and touch in others; these animals depend more than we do today on the sharpest possible awareness of a hostile environment. Apart from the complexity of the cerebral cortex (which does not directly perceive pain) their nervous systems are almost identical to ours and their reactions to pain remarkably similar, though lacking (so far as

we know) the philosophical and moral overtones. The emotional element is all too evident, mainly in the form of fear and anger.[9]

In Britain, three separate expert government committees on matters relating to animals have accepted the conclusion that animals feel pain. After noting the obvious behavioral evidence for this view, the members of the Committee on Cruelty to Wild Animals, set up in 1951, said:

> . . . we believe that the physiological, and more particularly the anatomical, evidence fully justifies and reinforces the commonsense belief that animals feel pain.

And after discussing the evolutionary value of pain the committee's report concluded that pain is "of clear-cut biological usefulness" and this is "a third type of evidence that animals feel pain." The committee members then went on to consider forms of suffering other than mere physical pain and added that they were "satisfied that animals do suffer from acute fear and terror." Subsequent reports by British government committees on experiments on animals and on the welfare of animals under intensive farming methods agreed with this view, concluding that animals are capable of suffering both from straightforward physical injuries and from fear, anxiety, stress, and so on.[10] Finally, within the last decade, the publication of scientific studies with titles such as *Animal Thought, Animal Thinking,* and *Animal Suffering: The Science of Animal Welfare* have made it plain that conscious awareness in nonhuman animals is now generally accepted as a serious subject for investigation.[11]

That might well be thought enough to settle the matter; but one more objection needs to be considered. Human beings in pain, after all, have one behavioral sign that nonhuman animals do not have: a developed language. Other animals may communicate with each other, but not, it seems, in the complicated way we do. Some philosophers, including Descartes, have thought it important that while humans can tell each other about their experience of pain in great detail, other animals cannot. (Interestingly, this once neat dividing line between humans and other species has now been threatened by the discovery that chimpanzees can be taught a language.[12]) But as Bentham pointed out long ago, the ability to use language is not relevant to the question of how a being ought to be treated—unless that ability can be linked to the capacity to suffer, so that the absence of a language casts doubt on the existence of this capacity.

This link may be attempted in two ways. First, there is a hazy line of philosophical thought, deriving perhaps from some doctrines associated with the influential philosopher Ludwig Wittgenstein, which maintains that we cannot meaningfully attribute states of consciousness to beings without language. This position seems to me very implausible. Language may be necessary for abstract thought, at some level anyway; but states like pain are more primitive, and have nothing to do with language.

The second and more easily understood way of linking language and the existence of pain is to say that the best evidence we can have that other creatures are in pain is that they tell us that they are. This is a distinct line of argument, for it is denying not that non–language-users conceivably *could* suffer, but only that we could ever have sufficient reason to *believe* that they are suffering. Still, this line of argument fails too. As Jane Goodall has pointed out in her study of chimpanzees, *In the Shadow of Man*, when it comes to the expression of feelings and emotions language is less important than nonlinguistic modes of communication such as a cheering pat on the back, an exuberant embrace, a clasp of the hands, and so on. The basic signals we use to convey pain, fear, anger, love, joy, surprise, sexual arousal, and many other emotional states are not specific to our own species.[13] The statement "I am in pain" may be one piece of evidence for the conclusion that the speaker is in pain, but it is not the only possible evidence, and since people sometimes tell lies, not even the best possible evidence.

Even if there were stronger grounds for refusing to attribute pain to those who do not have a language, the consequences of this refusal might lead us to reject the conclusion. Human infants and young children are unable to use language. Are we to deny that a year-old child can suffer? If not, language cannot be crucial. Of course, most parents understand the responses of their children better than they understand the responses of other animals; but this is just a fact about the relatively greater knowledge that we have of our own species and the greater contact we have with infants as compared to animals. Those who have studied the behavior of other animals and those who have animals as companions soon learn to understand their responses as well as we understand those of an infant, and sometimes better.

So to conclude: there are no good reasons, scientific or philosophical, for denying that animals feel pain. If we do not doubt that other humans feel pain we should not doubt that other animals do so too.

Animals can feel pain. As we saw earlier, there can be no moral justification for regarding the pain (or pleasure) that animals feel as less important than the same amount of pain (or pleasure) felt by humans. But what practical consequences follow from this conclusion? To prevent misunderstanding I shall spell out what I mean a little more fully.

If I give a horse a hard slap across its rump with my open hand, the horse may start, but it presumably feels little pain. Its skin is thick enough to protect it against a mere slap. If I slap a baby in the same way, however, the baby will cry and presumably feel pain, for its skin is more sensitive. So it is worse to slap a baby than a horse, if both slaps are administered with equal force. But there must be some kind of blow—I don't know exactly what it would be, but perhaps a blow with a heavy stick—that would cause the horse as much pain as we cause a baby by slapping it with our hand. That is what I mean by "the same amount of pain," and if we consider it wrong to inflict that much pain on a baby for no good reason then we must, unless we are speciesists, consider it equally wrong to inflict the same amount of pain on a horse for no good reason.

Other differences between humans and animals cause other complications. Normal adult human beings have mental capacities that will, in certain circumstances, lead them to suffer more than animals would in the same circumstances. If, for instance, we decided to perform extremely painful or lethal scientific experiments on normal adult humans, kidnapped at random from public parks for this purpose, adults who enjoy strolling in parks would become fearful that they would be kidnapped. The resultant terror would be a form of suffering additional to the pain of the experiment. The same experiments performed on nonhuman animals would cause less suffering since the animals would not have the anticipatory dread of being kidnapped and experimented upon. This does not mean, of course, that it would be *right* to perform the experiment on animals, but only that there is a reason, which is *not* speciesist, for preferring to use animals rather than normal adult human beings, if the experiment is to be done at all. It should be noted, however, that this same argument gives us a reason for preferring to use human infants—orphans perhaps—or severely retarded human beings for experiments, rather than adults, since infants and retarded humans would also have no idea of what was going to happen to them. So far as this argument is concerned nonhuman animals and infants and retarded humans are in the same category; and if we use this argument to justify experiments on nonhuman animals we have to ask ourselves whether we are also prepared to allow experiments on human infants and retarded adults; and if we make a distinction between animals and these humans, on what basis can we do it, other than a bare-faced—and morally indefensible—preference for members of our own species?

There are many matters in which the superior mental powers of normal adult humans make a difference: anticipation, more detailed memory, greater knowledge of what is happening, and so on. Yet these differences do not all point to greater suffering on the part of the normal human being. Sometimes animals may suffer more because of their more limited understanding. If, for instance, we are taking prisoners in wartime we can explain to them that although they must submit to capture, search, and confinement, they will not otherwise be harmed and will be set free at the conclusion of hostilities. If we capture wild animals, however, we cannot explain that we are not threatening their lives. A wild animal cannot distinguish an attempt to overpower and confine from an attempt to kill; the one causes as much terror as the other.

It may be objected that comparisons of the sufferings of different species are impossible to make and that for this reason when the interests of animals and humans clash the principle of equality gives no guidance. It is probably true that comparisons of suffering between members of different species cannot be made precisely, but precision is not essential. Even if we were to prevent the infliction of suffering on animals only when it is quite certain that the interests of humans will not be affected to anything like the extent that animals are affected, we would be forced to make radical changes in our treatment of animals that would involve our diet, the farming methods we use, experimental procedures in many fields of science, our approach to wildlife and to hunting, trapping and the wearing of furs, and areas of entertainment like circuses, rodeos, and zoos. As a result, a vast amount of suffering would be avoided.

So far I have said a lot about inflicting suffering on animals, but nothing about killing them. This omission has been deliberate. The application of the principle of equality to the infliction of suffering is, in theory at least, fairly straightforward. Pain and suffering are in themselves bad and should be prevented or minimized, irrespective of the race, sex, or species of the being that suffers. How bad a pain is depends on how intense it is and how long it lasts, but pains of the same intensity and duration are equally bad, whether felt by humans or animals.

The wrongness of killing a being is more complicated. I have kept, and shall continue to keep, the question of killing in the background because in the present state of human tyranny over other species the more simple, straightforward principle of equal consideration of pain or pleasure is a sufficient basis for identifying and protesting against all the major abuses of animals that human beings practice. Nevertheless, it is necessary to say something about killing.

Just as most human beings are speciesists in their readiness to cause pain to animals when they would not cause a similar pain to humans for the same reason, so most human beings are speciesists in their readiness to kill other animals when they would not kill human beings. We need to proceed more cautiously here, however, because people hold widely differing views about when it is legitimate to kill humans, as the continuing debates over abortion and euthanasia attest. Nor have moral philosophers been able to agree on exactly what it is that makes it wrong to kill human beings, and under what circumstances killing a human being may be justifiable.

Let us consider first the view that it is always wrong to take an innocent human life. We may call this the "sanctity of life" view. People who take this view oppose abortion and euthanasia. They do not usually, however, oppose the killing of nonhuman animals—so perhaps it would be more accurate to describe this view as the "sanctity of *human* life" view. The belief that human life, and only human life, is sacrosanct is a form of speciesism. To see this, consider the following example.

Assume that, as sometimes happens, an infant has been born with massive and irreparable brain damage. The damage is so severe that the infant can never be any more than a "human vegetable," unable to talk, recognize other people, act independently of others, or develop a sense of self-awareness. The parents of the infant, realizing that they cannot hope for any improvement in their child's condition and being in any case unwilling to spend, or ask the state to spend, the thousands of dollars that would be needed annually for proper care of the infant, ask the doctor to kill the infant painlessly.

Should the doctor do what the parents ask? Legally, the doctor should not, and in this respect the law reflects the sanctity of life view. The life of every human being is sacred. Yet people who would say this about the infant do not object to the killing of nonhuman animals. How can they justify their different judgments? Adult chimpanzees, dogs, pigs, and members of many other species far surpass the brain-damaged infant in their ability to relate to others, act independently, be self-aware, and any other capacity that could reasonably be said to give value to life. With the most intensive care possible, some severely retarded infants can never achieve the intelligence level of a dog. Nor can we appeal to the concern of the infant's parents, since they themselves, in this imaginary example (and in some actual cases) do not want the infant kept alive. The only thing that distinguishes the infant from the animal, in the eyes of those who claim it has a "right to life," is that it is, biologically, a member of the species Homo sapiens, whereas chimpanzees, dogs, and pigs are not. But to use *this* difference as the basis for granting a right to life to the infant and not to the other animals is, of course, pure speciesism.[14] It is exactly the kind of arbitrary difference that the most crude and overt kind of racist uses in attempting to justify racial discrimination.

This does not mean that to avoid speciesism we must hold that it is as wrong to kill a dog as it is to kill a human being in full possession of his or her faculties. The only position that is irredeemably speciesist is the one that tries to make the boundary of the right to life run exactly parallel to the boundary of our own species. Those who hold the sanctity of life view do this, because while distinguishing sharply between human beings and other animals they allow no distinctions to be made within our own species, objecting to the killing of the severely retarded and the hopelessly senile as strongly as they object to the killing of normal adults.

To avoid speciesism we must allow that beings who are similar in all relevant respects have a similar right to life—and mere membership in our own biological species cannot be a morally relevant criterion for this right. Within these limits we could still hold, for instance, that it is worse to kill a normal adult human, with a capacity for self-awareness and the ability to plan for the future and have meaningful relations with others, than it is to kill a mouse, which presumably does not share all of these characteristics; or we might appeal to the close family and other personal ties that humans have but mice do not have to the same degree; or we might think that it is the consequences for other humans, who will be put in fear for their own lives, that makes the crucial difference; or we might think it is some combination of these factors, or other factors altogether.

Whatever criteria we choose, however, we will have to admit that they do not follow precisely the boundary of our own species. We may legitimately hold that there are some features of certain beings that make their lives more valuable than those of other beings; but there will surely be some nonhuman animals whose lives, by any standards, are more valuable than the lives of some humans. A chimpanzee, dog, or pig, for instance, will have a higher degree of self-awareness and a greater capacity for meaningful relations with others than a severely retarded infant or someone in a state of advanced senility. So if we base the right to life on these characteristics we must grant these animals a right to life as good as, or better than, such retarded or senile humans.

This argument cuts both ways. It could be taken as showing that chimpanzees, dogs, and pigs, along with some other species, have a right to life and we commit a grave moral offense whenever we kill them, even when they are old and suffering and our intention is to put them out of their misery. Alternatively one could take the argument as showing that the severely retarded and hopelessly senile have no right to life and may be killed for quite trivial reasons, as we now kill animals.

Since the main concern of this book is with ethical questions having to do with animals and not with the morality of euthanasia I shall not attempt to settle this issue finally.[15] I think it is reasonably clear, though, that while both of the positions just described avoid speciesism, neither is satisfactory. What we need is some middle position that would avoid speciesism but would not make the lives of the retarded and senile as cheap as the lives of pigs and dogs now are, or make the lives of pigs and dogs so sacrosanct that we think it wrong to put them out of hopeless misery. What we must do is bring nonhuman animals within our sphere of moral concern and cease to treat their lives as expendable for whatever trivial purposes we may have. At the same time, once we realize that the fact that a being is a member of our own species is not in itself enough to make it always wrong to kill that being, we may come to reconsider our policy of preserving human lives at all costs, even when there is no prospect of a meaningful life or of existence without terrible pain.

I conclude, then, that a rejection of speciesism does not imply that all lives are of equal worth. While self-awareness, the capacity to think ahead and have hopes and aspirations for the future, the capacity for meaningful relations with others and so on are not relevant to the question of inflicting pain—since pain is pain, whatever other capacities, beyond the capacity to feel pain, the being may have—these capacities are relevant to the question of taking life. It is not arbitrary to hold that the life of a self-aware being, capable of abstract thought, of planning for the future, of complex acts of communication, and so on, is more valuable than the life of a being without these capacities. To see the difference between the issues of inflicting pain and taking life, consider how we would choose within our own species. If we had to choose to save the life of a normal human being or an intellectually disabled human being, we would probably choose to save the life of a normal human being; but if we had to choose between preventing pain in the normal human being or the intellectually disabled one—imagine that both have received painful but superficial injuries, and we only have enough painkiller for one of them—it is not nearly so clear how we ought to choose. The same is true when we consider other species. The evil of pain is, in itself, unaffected by the other characteristics of the being who feels the pain; the value of life is affected by these other characteristics. To give just one reason for this difference, to take the life of a being who has been hoping, planning, and working for some future goal is to deprive that being of the fulfillment of all those efforts; to take the life of a being with a mental capacity below the level needed to grasp that one is a being with a future—much less make plans for the future—cannot involve this particular kind of loss.[16]

Normally this will mean that if we have to choose between the life of a human being and the life of another animal we should choose to save the life of the human; but there may be special cases in which the reverse holds true, because the human being in question does not have the capacities of a normal human being. So this view is not speciesist, although it may appear to be at first glance. The preference, in normal cases, for saving a human life over the life of an animal when a choice *has* to be made is a preference based on the characteristics that normal humans have, and not on the mere fact that they are members of our own species. This is why when we consider members of our own species who lack the characteristics of normal humans we can no longer say that their lives are always to be preferred to those of other animals. This issue comes up in a practical way in the following chapter. In general, though, the question of when it is wrong to kill (painlessly) an animal is one to which we need give no precise answer. As long as we remember that we should give the same respect to the lives of animals as we give to the lives of those humans at a similar mental level, we shall not go far wrong.[17]

In any case, the conclusions that are argued for in this book flow from the principle of minimizing suffering alone. The idea that it is also wrong to kill animals painlessly gives some of these conclusions additional support that is welcome but strictly unnecessary. Interestingly enough, this is true even of the conclusion that we ought to become vegetarians, a conclusion that in the popular mind is generally based on some kind of absolute prohibition on killing.

The reader may already have thought of some objections to the position I have taken in this chapter. What, for instance, do I propose to do about animals who may cause harm to human beings? Should we try to stop animals from killing each other? How do we know that plants cannot feel pain, and if they can, must we starve? To avoid interrupting the flow of the main argument I have chosen to discuss these and other objections in a separate chapter, and readers who are impatient to have their objections answered may look ahead to Chapter 6.

The next two chapters explore two examples of speciesism in practice. I have limited myself to two examples so that I would have space for a reasonably thorough discussion, although this limit means that the book contains no discussion at all of other practices that exist only because we do not take seriously the interests of other animals—practices like hunting, whether for sport or for furs; farming minks, foxes, and other animals for their fur; capturing wild animals (often after shooting their mothers) and imprisoning them in small cages for humans to stare at; tormenting animals to make them learn tricks for circuses and tormenting them to make them entertain the audiences at rodeos; slaughtering whales with explosive harpoons, under the guise of scientific research; drowning over 100,000 dolphins annually in nets set by tuna fishing boats; shooting three million kangaroos every year in the Australian outback to turn them into skins and pet food; and generally ignoring the interests of wild animals as we extend our empire of concrete and pollution over the surface of the globe.

I shall have nothing, or virtually nothing, to say about these things, because as I indicated in the preface to this edition, this book is not a compendium of all the nasty things we do to animals. Instead I have chosen two central illustrations of speciesism in practice. They are not isolated examples of sadism, but practices that involve, in one case, tens of millions of animals, and in the other, billions of animals every year. Nor can we pretend that we have nothing to do with these practices. One of them—experimentation on animals—is promoted by the government we elect and is largely paid for out of the taxes we pay. The other—rearing animals for food—is possible only because most people buy and eat the products of this practice. That is why I have chosen to discuss these particular forms of speciesism. They are at its heart. They cause more suffering to a greater number of animals than anything else that human beings do. To stop them we must change the policies of our government, and we must change our own lives, to the extent of changing our diet. If these officially promoted and almost universally accepted forms of speciesism can be abolished, abolition of the other speciesist practices cannot be far behind.

Chapter 2

Tools for Research . . .

your taxes at work

Project X, a popular film released in 1987, gave many Americans their first glimpse into animal experiments carried out by their own armed forces. The film's plot centers on an air force experiment designed to see whether chimpanzees could continue to "fly" a simulated plane after being exposed to radiation. A young air force cadet assigned to duty in the laboratory becomes attached to one particular chimpanzee, with whom he can communicate in sign language. When this chimpanzee's turn for exposure to radiation comes, the young man (with the assistance of his attractive girlfriend, naturally) determines to liberate the chimpanzees.

The plot was fiction, but the experiments were not. They were based on experiments that have been conducted over many years at Brooks Air Force Base, in Texas, and variations of which are continuing. But filmgoers did not get the whole story. What happened to the chimpanzees in the film was very much a softened version of what really happens. So we should consider the experiments themselves, as described in documents issued by Brooks Air Force Base.

As indicated in the film, the experiments involve a kind of flight simulator. The device is known as a Primate Equilibrium Platform, or PEP. It consists of a platform that can be made to pitch and roll like an airplane. The monkeys sit in a chair that is part of the platform. In front of them is a control stick, by means of which the platform can be returned to a horizontal position. Once monkeys have been trained to do this, they are subjected to radiation and to chemical warfare agents, to see how these affect their ability to fly. (A photograph of the Primate Equilibrium Platform appears following page 157.)

The standard training procedure for the PEP is described in a Brooks Air Force Base publication entitled "Training Procedure for Primate Equilibrium Platform."[1] The following is a summary:

Phase I (chair adaptation): The monkeys are "restrained" (in other words, tied down) in the PEP chair for one hour per day for five days, until they sit quietly.

Phase II (stick adaptation): The monkeys are restrained in the PEP chair. The chair is then tipped forward and the monkeys are given electric shocks. This causes the monkey to "turn in the chair or bite the platform. . . . This behavior is redirected toward the [experimenter's] gloved hand which is placed directly over the control stick." Touching the hand results in the shock being stopped, and the monkey (who has not been fed that day) is given a raisin. This happens to each monkey one hundred times a day for between five and eight days.

Phase III (stick manipulation): This time when the PEP is tipped forward, merely touching the stick is not enough to stop the electric shock. The monkeys continue to receive electric shocks until they pull the stick back. This is repeated one hundred times per day.

Phases IV–VI (push stick forward and pull stick back): In these phases the PEP is tipped back and the monkeys are shocked until they push the stick forward. Then the PEP is again tilted forward, and they must again learn to pull the stick back. This is repeated one hundred times per day. Then the platform switches randomly between backward and forward and the monkeys are again shocked until they make the appropriate response.

Phase VII (control stick operational): Up to this point, although the monkeys have been pulling the control stick backward and forward, it has not affected the position of the platform. Now the monkey controls the position of the platform by pulling the stick. In this phase the automatic shocker does not function. Shocks are manually given at approximately every three or four seconds for a 0.5 second duration. This is a slower rate than previously, to ensure that correct behavior is not punished and therefore, to use the jargon of the manual, "extinguished." If the monkey does stop performing as desired, the training returns to phase VI. Otherwise, training continues in this phase until the monkey can maintain the platform at a nearly horizontal level and avoid 80 percent of the shocks given.

The time taken for training the monkeys in phases III through VII is ten to twelve days.

After this period, training continues for another twenty days. During this further period a randomizing device is used to make the chair pitch and roll more violently, but the monkey must maintain the same level of performance in returning the chair to the horizontal or else receive frequent electric shocks.

All this training, involving thousands of electric shocks, is only preliminary to the real experiment. Once the monkeys are regularly keeping the platform horizontal most of the time, they are exposed to lethal or sublethal doses of radiation or to chemical warfare agents, to see how long they can continue to "fly" the platform. Thus, nauseous and probably vomiting from a fatal dose of radiation, they are forced to try to keep the platform horizontal, and if they fail they receive frequent electric shocks. Here is one example, taken from a United States Air Force School of Aerospace Medicine report published in October 1987—after *Project X* had been released.[2]

The report is entitled "Primate Equilibrium Performance Following Soman Exposure: Effects of Repeated Daily Exposures to Low Soman Doses." Soman is another name for nerve gas, a chemical warfare agent that caused terrible agony to troops in the First World War, but fortunately has been very little used in warfare since then. The report begins by referring to several previous reports in which the same team of investigators studied the effects of "acute exposure to soman" on performance in the Primate Equilibrium Platform. This particular study, however, is on the effect of low doses received over several days. The monkeys in this experiment had been operating the platform "at least weekly" for a minimum of two years and had received various drugs and low doses of soman before, but not within the previous six weeks.

The experimenters calculated the doses of soman that would be sufficient to reduce the monkeys' ability to operate the platform. For the calculation to be made, of course, the monkeys would have been receiving electric shocks because of their inability to keep the platform level. Although the report is mostly concerned with the effect of the nerve poison on the performance level of the monkeys, it does give some insight into other effects of chemical weapons:

> The subject was completely incapacitated on the day following the last exposure, displaying neurological symptoms including gross incoordination, weakness, and intention tremor.... These symptoms persisted for several days, during which the animal remained unable to perform the PEP task.[3]

Dr. Donald Barnes was for several years principal investigator at the U.S. Air Force School of Aerospace Medicine, and in charge of the experiments with the Primate Equilibrium Platform at Brooks Air Force Base. Barnes estimates that he irradiated about one thousand trained monkeys during his years in this position. Subsequently he has written:

> For some years, I had entertained suspicions about the utility of the data we were gathering. I made a few token attempts to ascertain both the destination and the purpose of the technical reports we published but now acknowledge my eagerness to accept assurances from those in command that we were, in fact, providing a real service to the U.S. Air Force and, hence, to the defense of the free world. I used those assurances as blinkers to avoid the reality of what I saw in the field, and even though I did not always wear them comfortably, they did serve to protect me from the insecurities associated with the potential loss of status and income....
>
> And then, one day, the blinkers slipped off, and I found myself in a very serious confrontation with Dr. Roy DeHart, Commander, U.S. Air Force School of Aerospace Medicine. I tried to point out that, given a nuclear confrontation, it is highly unlikely that operational commanders will go to charts and figures based upon data from the rhesus monkey to gain estimates of probable force strength or second strike capability. Dr. DeHart insisted that the data will be invaluable, asserting, "They don't know the data are based on animal studies."[4]

Barnes resigned and has become a strong opponent of animal experimentation; but experiments using the Primate Equilibrium Platform have continued.

Project X lifted the veil on one kind of experiment conducted by the military. We have now examined that in a little detail, although it would take a long time to describe all the forms of radiation and chemical warfare agents tested, in varying doses, on monkeys in the Primate Equilibrium Platform. What we now need to grasp is that this is just one very small part of the total amount of military experimentation on animals. Concern about this experimentation goes back several years.

In July 1973 Representative Les Aspin of Wisconsin learned through an advertisement in an obscure newspaper that the United States Air Force was planning to purchase two hundred beagle puppies, with vocal cords tied to prevent normal barking, for tests of poisonous gases. Shortly afterward it became known that the army was also proposing to use beagles—four hundred this time—in similar tests.

Aspin began a vigorous protest, supported by antivivisection societies. Advertisements were placed in major newspapers across the country. Letters from an outraged public began pouring in. An aide from the House of Representatives Armed Services Committee said that the committee had received more mail on the beagles than it had received on any other subject since Truman sacked General MacArthur, while an internal Department of Defense memo released by Aspin said that the volume of mail the department had received was the greatest ever for any single event, surpassing even the mail on the bombings of North Vietnam and Cambodia.[5] After defending the experiments initially, the Defense Department then announced that it was postponing them and looking into the possibility of replacing the beagles with other experimental animals.

All this amounted to a curious incident—curious because the public furor over this particular experiment implied a remarkable ignorance of the nature of standard experiments performed by the armed services, research establishments, universities, and commercial firms of many different kinds. True, the proposed air force and army experiments were designed so that many animals would suffer and die without any certainty that this suffering and death would save a single human life or benefit humans in any way at all; but the same can be said of millions of other experiments performed each year in the United States alone. Perhaps the concern arose because the experiments were to be done on beagles. But if so, why has there been no protest at the following experiment, conducted more recently:

Under the direction of the U.S. Army Medical Bioengineering Research and Development Laboratory at Fort Detrick, in Frederick, Maryland, researchers fed 60 beagle dogs varied doses of the explosive TNT. The dogs were given the TNT in capsules every day for six months. Symptoms observed included dehydration, emaciation, anemia, jaundice, low body temperature, discolored urine and feces, diarrhea, loss of appetite and weight loss, enlarged livers, kidneys and spleen, and the beagles became uncoordinated. One female was "found to be moribund [dying]" during week 14 and was killed; another was found dead during week 16. The report states that the experiment represents "a portion" of the data which the Fort Detrick laboratory is developing on the effects of TNT on mammals. Because injuries were observed even at the lowest doses, the study failed to establish the level at which TNT had no observable effects; thus, the report concludes "additional studies . . . of TNT in beagle dogs may be warranted."[6]

In any case, it is wrong to limit our concern to dogs. People tend to care about dogs because they generally have more experience with dogs as companions; but other animals are as capable of suffering as dogs are. Few people feel sympathy for rats. Yet rats are intelligent animals, and there can be no doubt that rats are capable of suffering and do suffer from the countless painful experiments performed on them. If the army were to stop experimenting on dogs and switch to rats instead, we should not be any less concerned.

Some of the worst military experiments are carried out at a place known as AFRRI—the Armed Forces Radiobiology Research Institute, in Bethesda, Maryland. Here, instead of using a Primate Equilibrium Platform, experimenters have tied animals down in chairs and irradiated them or have trained them to press levers and observed the effects of irradiation on their performance. They have also trained monkeys to run in an "activity wheel," which is a kind of cylindrical treadmill. (See photograph following page 157.) The monkeys receive electric shocks unless they keep the wheel moving at speeds above one mile per hour.

In one experiment using the primate activity wheel, Carol Franz of the behavioral sciences department at AFRRI trained thirty-nine monkeys for nine weeks, two hours per day, until they could alternate "work" and "rest" periods for six continuous hours. They were then subjected to varying doses of radiation. Monkeys receiving the higher doses vomited up to seven times. They were then put back into the activity wheel to measure the effect of the radiation on their ability to "work." During this period, if a monkey did not move the wheel for one minute, "shock intensity was increased to 10 mA." (This is an extremely intense electric shock, even by the quite excessive standards of American animal experimentation; it must cause very severe pain.) Some monkeys continued to vomit while in the activity wheel. Franz reports the effect that the various doses of radiation had on performance. The report also indicates that the irradiated monkeys took between a day and a half and five days to die.[7]

Since I do not wish to spend this entire chapter describing experiments conducted by the United States armed forces, I shall turn now to nonmilitary experimentation (although we shall, in passing, examine one or two other military experiments where they are relevant to other topics). Meanwhile, I hope that United States taxpayers, whatever they think the size of the military budget should be, will ask themselves: Is this what I want the armed forces to be doing with my taxes?

We should not, of course, judge all animal experimentation by the experiments I have just described. The armed services, one might think, are hardened to suffering by their concentration on war, death, and injury. Genuine scientific research, surely, will be very different, won't it? We shall see. To begin our examination of nonmilitary scientific research, I shall allow Professor Harry F. Harlow to speak for himself. Professor Harlow, who worked at the Primate Research Center in Madison, Wisconsin, was for many years editor of a leading psychology journal, and until his death a few years ago was held in high esteem by his colleagues in psychological research. His work has been cited approvingly in many basic textbooks of psychology, read by millions of students taking introductory psychology courses over the last twenty years. The line of research he began has been continued after his death by his associates and former students.

In a 1965 paper, Harlow describes his work as follows:

> For the past ten years we have studied the effects of partial social isolation by raising monkeys from birth onwards in

> bare wire cages.... These monkeys suffer total maternal deprivation.... More recently we have initiated a series of studies on the effects of total social isolation by rearing monkeys from a few hours after birth until 3, 6, or 12 months of age in [a] stainless steel chamber. During the prescribed sentence in this apparatus the monkey has no contact with any animal, human or sub-human.

These studies, Harlow continues, found that

> sufficiently severe and enduring early isolation reduces these animals to a social-emotional level in which the primary social responsiveness is fear.[8]

In another article Harlow and his former student and associate Stephen Suomi described how they were trying to induce psychopathology in infant monkeys by a technique that appeared not to be working. They were then visited by John Bowlby, a British psychiatrist. According to Harlow's account, Bowlby listened to the story of their troubles and then toured the Wisconsin laboratory. After he had seen the monkeys individually housed in bare wire cages he asked, "Why are you trying to produce psychopathology in monkeys? You already have more psychopathological monkeys in the laboratory than have ever been seen on the face of the earth."[9]

Bowlby, incidentally, was a leading researcher on the consequences of maternal deprivation, but his research was conducted with children, primarily war orphans, refugees, and institutionalized children. As far back as 1951, before Harlow even began his research on nonhuman primates, Bowlby concluded:

> The evidence has been reviewed. It is submitted that evidence is now such that it leaves no room for doubt regarding the general proposition that the prolonged deprivation of the young child of maternal care may have grave and far-reaching effects on his character and so on the whole of his future life.[10]

This did not deter Harlow and his colleagues from devising and carrying out their monkey experiments.

In the same article in which they tell of Bowlby's visit, Harlow and Suomi describe how they had the "fascinating idea" of inducing depression by "allowing baby monkeys to attach to cloth surrogate mothers who could become monsters":

> The first of these monsters was a cloth monkey mother who, upon schedule or demand, would eject high-pressure compressed air. It would blow the animal's skin practically off its body. What did the baby monkey do? It simply clung tighter and tighter to the mother, because a frightened infant clings to its mother at all costs. We did not achieve any psychopathology.
>
> However, we did not give up. We built another surrogate monster mother that would rock so violently that the baby's head and teeth would rattle. All the baby did was cling tighter and tighter to the surrogate. The third monster we built had an embedded wire frame within its body which would spring forward and eject the infant from its ventral surface. The infant would subsequently pick itself off the floor, wait for the frame to return into the cloth body, and then cling again to the surrogate. Finally, we built our porcupine mother. On command, this mother would eject sharp brass spikes over all of the ventral surface of its body. Although the infants were distressed by these pointed rebuffs, they simply waited until the spikes receded and then returned and clung to the mother.

These results, the experimenters remark, were not so surprising, since the only recourse of an injured child is to cling to its mother.

Eventually, Harlow and Suomi gave up on the artificial monster mothers because they found something better: a real monkey mother who was a monster. To produce such mothers, they reared female monkeys in isolation, and then tried to make them pregnant. Unfortunately the females did not have normal sexual relations with male monkeys, so they had to be made pregnant by a technique that Harlow and Suomi refer to as a "rape rack." When the babies were born the experimenters observed the monkeys. They found that some simply ignored the infants, failing to cuddle the crying baby to the breast as normal monkeys do when they hear their baby cry. The other pattern of behavior observed was different:

> The other monkeys were brutal or lethal. One of their favorite tricks was to crush the infant's skull with their teeth. But the really sickening behavior pattern was that of smashing the infant's face to the floor, and then rubbing it back and forth.[11]

In a 1972 paper, Harlow and Suomi say that because depression in humans has been characterized as embodying a state of "helplessness and hopelessness, sunken in a well of despair," they designed a device "on an intuitive basis" to reproduce such a "well of despair" both physically and psychologically. They built a vertical chamber with stainless steel sides sloping inward to form a rounded bottom and placed a young monkey in it for periods of up to forty-five days. They found that after a few days of this confinement the monkeys "spend most of their time huddled in a corner of the chamber." The confinement produced "severe and persistent psychopathological behavior of a depressive nature." Even nine months after release the monkeys would sit clasping their arms around their bodies instead of moving around and exploring their surroundings as normal monkeys do. But the report ends inconclusively and ominously:

> Whether [the results] can be traced specifically to variables such as chamber shape, chamber size, duration of confinement, age at time of confinement or, more likely, to a combination of these and other variables remains the subject of further research.[12]

Another paper explains how, in addition to the "well of despair," Harlow and his colleagues created a "tunnel of terror" to produce terrified monkeys,[13] and in yet another report Harlow describes how he was able "to induce psychological death in rhesus monkeys" by providing them with terry cloth–covered "mother surrogates" that were normally kept at a temperature of 99 degrees Fahrenheit, but could be rapidly chilled to 35 degrees Fahrenheit to simulate a kind of maternal rejection.[14]

Harlow is now dead, but his students and admirers have spread across the United States and continue to perform experiments in a similar vein. John P. Capitanio, under the direction of one of Harlow's students, W. A. Mason, has conducted deprivation experiments at the California Primate Research Center at the University of California, Davis. In these experiments, Capitanio compared the social behavior of rhesus monkeys "reared" by a dog with that of monkeys "reared" by a plastic hobbyhorse. He concluded that "although members of both groups were clearly abnormal in the extent of their social interactions," the monkeys who had been kept with the dog coped better than those kept with the plastic toy.[15]

After leaving Wisconsin, Gene Sackett continued deprivation studies at the University of Washington Primate Center. Sackett has raised rhesus monkeys, pigtail macaques, and crab-eating macaques in total isolation to study the differences in personal behavior, social behavior, and exploration behavior. He found differences among the different monkey species which "question the generality of the 'isolation syndrome' across primate species." If there are differences even among closely related species of monkeys, generalization from monkeys to humans must be far more questionable.[16]

Martin Reite of the University of Colorado conducted deprivation experiments on bonnet monkeys and pigtailed macaques. He was aware that Jane Goodall's observations of orphaned wild chimpanzees described "profound behavioral disturbances, with sadness or depressive affective changes as major components." But because "in comparison with monkey studies, relatively little has been published on experimental separations in great apes," he and other experimenters decided to study seven infant chimpanzees who had been separated from their mothers at birth and reared in a nursery environment. After periods ranging between seven and ten months, some of the infants were placed in isolation chambers for five days. The isolated infants screamed, rocked, and threw themselves at the walls of the chamber. Reite concluded that "isolation in infant chimpanzees may be accompanied by marked behavioral changes" but noted that (you guessed it) more research was needed.[17]

Since Harlow began his maternal deprivation experiments some thirty years ago, over 250 such experiments have been conducted in the United States. These experiments subjected over seven thousand animals to procedures that induced distress, despair, anxiety, general psychological devastation, and death. As some of the preceding quotations show, research now feeds off itself. Reite and his colleagues experimented on chimpanzees because relatively little experimental work had been done on the great apes, as compared with monkeys. They apparently felt no need to address the basic question of why we should be doing any experiments on maternal deprivation in animals at all. They did not even try to justify their experiments by claiming they were of benefit to human beings. That we already have extensive observations of orphaned chimpanzees in the wild seems not to have been of interest to them. Their attitude was plain: this has been done with animals of one species, but not with animals of another, so let's do it to them. The same attitude recurs constantly throughout the psychological and behavioral sciences. The most amazing part of the story is that taxpayers have paid for all this research—to the tune of over $58 million for maternal deprivation research alone.[18] In this respect, but not only in this respect, animal experimentation in civilian life is not so different from military experimentation.

The practice of experimenting on nonhuman animals as it exists today throughout the world reveals the consequences of speciesism. Many experiments inflict severe pain without the remotest prospect of significant benefits for human beings or any other animals. Such experiments are not isolated instances, but part of a major industry. In Britain, where experimenters are required to report the number of "scientific procedures" performed on animals, official government figures show that 3.5 million scientific procedures were performed on animals in 1988.[19] In the United States there are no figures of comparable accuracy. Under the Animal Welfare Act, the U.S. secretary of agriculture publishes a report listing the number of animals used by facilities registered with it, but this is incomplete in many ways. It does not include rats, mice, birds, reptiles, frogs, or domestic farm animals used for experimental purposes; it does not include animals used in secondary schools; and it does not include experiments performed by facilities that do not transport animals interstate or receive grants or contracts from the federal government.

In 1986 the U.S. Congress Office of Technology Assessment (OTA) published a report entitled "Alternatives to Animal Use in Research, Testing and Education." The OTA researchers attempted to determine the number of animals used in experimentation in the U.S. and reported that "estimates of the animals used in the United States each year range from 10 million to upwards of 100 million." They concluded that the estimates were unreliable but their best guess was "at least 17 million to 22 million."[20]

This is an extremely conservative estimate. In testimony before Congress in 1966, the Laboratory Animal Breeders Association estimated that the number of mice, rats, guinea pigs, hamsters, and rabbits used for experimental purposes in 1965 was around 60 million.[21] In 1984 Dr. Andrew Rowan of Tufts University School of Veterinary Medicine estimated that approximately 71 million animals are used each year. In 1985 Rowan revised his estimates to distinguish between the number of animals produced, acquired, and actually used. This yielded an estimate of between 25 and 35 million animals used in experiments each year.[22] (This figure omits animals who die in shipping or are killed before the experiment begins.) A stock market analysis of just one major supplier of animals to laboratories, the Charles River Breeding Laboratory, stated that this company alone produced 22 million laboratory animals annually.[23]

The 1988 report issued by the Department of Agriculture listed 140,471 dogs, 42,271 cats, 51,641 primates, 431,457 guinea pigs, 331,945 hamsters, 459,254 rabbits, and 178,249 "wild animals": a total of 1,635,288 used in experimentation. Remember that this report does not bother to count rats and mice, and covers at most an estimated 10 percent of the total number of animals used. Of the nearly 1.6 million animals reported by the Department of Agriculture to have been used for experimental purposes, over 90,000 are reported to have experienced "unrelieved pain or distress." Again, this is probably at most 10 percent of the total number of animals suffering unrelieved pain and distress—and if experimenters are less concerned about causing unrelieved pain to rats and mice than they are to dogs, cats, and primates, it could be an even smaller proportion.

Other developed nations all use large numbers of animals. In Japan, for example, a very incomplete survey published in 1988 produced a total in excess of eight million.[24]

One way of grasping the nature of animal experimentation as a large-scale industry is to look at the commercial products to which it gives rise and the way in which they are sold. Among these "products" are, of course, the animals themselves. We have seen how many animals Charles River Breeding Laboratories produces. In journals like *Lab Animal,* animals are advertised as if they were cars. Beneath a photograph of two guinea pigs, one normal and one completely hairless, the advertising copy says:

> When it comes to guinea pigs, now you have a choice. You can opt for our standard model that comes complete with hair. Or try our new 1988 stripped down, hairless model for speed and efficiency.
>
> Our euthymic, hairless guinea pigs are the product of years of breeding. They can be used for dermatologic studies for hair producing agents. Skin sensitization. Transdermal therapy. Ultraviolet studies. And more.

An advertisement for Charles River in *Endocrinology* (June 1985) asked:

> "You want to see our operation?"
>
> When it comes to operations, we give you just what the doctor ordered. Hypophysectomies, adrenalectomies, castrations, thymectomies, ovariectomies and thyroidectomies. We perform thousands of "endocrinectomies" every month on rats, mice or hamsters. Plus additional special surgery (spleenectomy, nephrectomy, cecetomy) on request.... For surgically altered research animals to fit your very specific research animals needs, call [phone number]. Our operators are available almost any time.

In addition to the animals themselves, animal experiments have created a market for specialized equipment. *Nature,* a leading British scientific journal, carries a section called "New on the Market," which recently informed its readers about a new piece of research equipment:

> The latest animal research tool from Columbus Instruments is an air-tight animal treadmill for the collection of oxygen consumption data during exercise. The treadmill has isolated running lanes with separate electrical shock stimuli which can be configured for up to four rats or mice.... The

basic £9,737 system includes a belt speed controller and an adjustable voltage shocker. The £13,487 fully automatic system can be programmed to run consecutive experiments with rest periods in between, and automatically monitors the number of trips to the shocker grid, time spent running, and time spent on the shocker grid.[25]

Columbus Instruments make several other ingenious devices. In *Lab Animal* it advertises:

> The Columbus Instruments Convulsion Meter makes possible objective and quantitative measurements of animal convulsions. A sensor precision platform load cell converts the vertical components of convulsion force into proportional electrical signals.... The user must observe the animal's behavior and activate the meter by a push-button switch when a convulsion is noticed. At the end of the experiment the totalized force and the totalized time of the convulsions will be obtained.

Then there is *The Whole Rat Catalog*. Published by Harvard Bioscience, it consists of 140 pages of equipment for use in experimenting on small animals, all written in cute advertising jargon. Of the transparent plastic rabbit restrainers, for instance, the catalog tells us: "The only thing that wiggles is the nose!" Sometimes, however, a little sensitivity to the controversial nature of the subject is shown: the description of the Rodent Carrying Case suggests, "Use this unobtrusive case to carry your favorite animal from one place to another without attracting attention." In addition to the usual cages, electrodes, surgical implements, and syringes, the catalog advertises Rodent Restraint Cones, Harvard Swivel-Tether Systems, Radiation Resistant Gloves, Implantable FM Telemetry Equipment, Liquid Diets for Rats and Mice in Alcohol Studies, Decapitators for both small and large animals, and even a Rodent Emulsifier which "will quickly reduce the remains of a small animal to a homogenous suspension."[26]

Presumably corporations would not bother to manufacture and advertise such equipment unless they expected considerable sales. And the items are not going to be bought unless they are going to be used.

Among the tens of millions of experiments performed, only a few can possibly be regarded as contributing to important medical research. Huge numbers of animals are used in university departments such as forestry and psychology; many more are used for commercial purposes, to test new cosmetics, shampoos, food coloring agents, and other inessential items. All this can happen only because of our prejudice against taking seriously the suffering of a being who is not a member of our own species. Typically, defenders of experiments on animals do not deny that animals suffer. They cannot deny the animals' suffering, because they need to stress the similarities between humans and other animals in order to claim that their experiments may have some relevance for human purposes. The experimenter who forces rats to choose between starvation and electric shock to see if they develop ulcers (which they do) does so because the rat has a nervous system very similar to a human being's, and presumably feels an electric shock in a similar way.

There has been opposition to experimenting on animals for a long time. This opposition has made little headway because experimenters, backed by commercial firms that profit by supplying laboratory animals and equipment, have been able to convince legislators and the public that opposition comes from uninformed fanatics who consider the interests of animals more important than the interests of human beings. But to be opposed to what is going on now it is not necessary to insist that all animal experiments stop immediately. All we need to say is that experiments serving no direct and urgent purpose should stop immediately, and in the remaining fields of research, we should, whenever possible, seek to replace experiments that involve animals with alternative methods that do not.

To understand why this seemingly modest change would be so important we need to know more about the experiments that are now being performed and have been performed for a century. Then we will be able to assess the claim by defenders of the present situation that experiments on animals are done only for important purposes. The following pages, therefore, describe some experiments on animals. Reading the reports of these experiments is not a pleasant experience; but we have an obligation to inform ourselves about what is done in our own community, especially since we are paying, through our taxes, for most of this research. If the animals have to undergo these experiments, the least we can do is read the reports and inform ourselves about them. That is why I have not attempted to tone down or gloss over some of the things that are done to animals. At the same time I have not tried to make these things worse than they really are. The reports that follow are all drawn from accounts written by the experimenters themselves and published by them in the scientific journals in which experimenters communicate with one another.

Such accounts are inevitably more favorable to the experimenters than reports by an outside observer would be. There are two reasons for this. One is that the experimenters will not emphasize the suffering they have inflicted unless it is necessary to do so in order to communicate the results of the experiment, and this is rarely the case. Most suffering therefore goes unreported. Experimenters may consider it unnecessary to include in their reports any mention of what happens when electric shock devices are left on when they should have been turned off, when animals recover consciousness in the midst of an operation because of an improperly administered anesthetic, or when unattended animals sicken and die over the weekend. The second reason scientific journals are a source favorable to experimenters is that they include only those experiments that the experimenters and editors of the journals consider significant. A British government committee found that only about one quarter of experiments on animals ever found their way into print.[27] There is no reason to believe that accounts of a higher proportion of experiments are published in the United States; indeed since the proportion of minor colleges with researchers of lesser talents is much higher in the United States than in Britain, it seems probable that an even smaller proportion of experiments yield results of any significance at all.

So in reading the following pages bear in mind that they are drawn from sources favorable to the experimenters; and if the results of the experiments do not appear to be of sufficient importance to justify the suffering they caused, remember that these examples are all taken from the small fraction of experiments that editors considered significant enough to publish. One last warning. The reports published in the journals always appear under the names of the experimenters. I have generally retained these names, since I see no reason to protect experimenters behind a cloak of anonymity. Nevertheless, it should not be assumed that the people named are especially evil or cruel people. They are doing what they were trained to do and what thousands of their colleagues do. The experiments are intended to illustrate not sadism on the part of individual experimenters but the institutionalized mentality of speciesism that makes it possible for these experimenters to do these things without serious consideration of the interests of the animals they are using.

Many of the most painful experiments are performed in the field of psychology. To give some idea of the numbers of animals experimented on in psychology laboratories, consider that during 1986 the National Institute of Mental Health funded 350 experiments on animals. The NIMH is just one source of federal funding for psychological experimentation. The agency spent over $11 million on experiments that involved direct manipulation of the brain, over $5 million on experiments that studied the effects drugs have on behavior, almost $3 million on learning and memory experiments, and over $2 million on experiments involving sleep deprivation, stress, fear, and anxiety. This government agency spent more than $30 million dollars on animal experiments in one year.[28]

One of the most common ways of experimenting in the field of psychology is to apply electric shocks to animals. This may be done with the aim of finding out how animals react to various kinds of punishment or to train animals to perform different tasks. In the first edition of this book I described experiments conducted in the late Sixties and early Seventies in which experimenters gave electric shocks to animals. Here is just one example from that period:

O. S. Ray and R. J. Barrett, working in the psychology research unit of the Veterans Administration Hospital, Pittsburgh, gave electric shocks to the feet of 1,042 mice. They then caused convulsions by giving more intense shocks through cup-shaped electrodes applied to the animals' eyes or through clips attached to their ears. They reported that unfortunately some of the mice who "successfully completed Day One training were found sick or dead prior to testing on Day Two."[29]

Now, nearly twenty years later, as I write the second edition of this book, experimenters are still dreaming up trifling new variations to try out on animals: W. A. Hillex and M. R. Denny of the University of California at San Diego placed rats in a maze and gave them electric shocks if, after one incorrect choice, on their next trial they failed to choose which way to go within three seconds. They concluded that the "results are clearly reminiscent of the early work on fixation and regression in the rat, in which the animals were typically shocked in the stem of the T-maze just preceding the choice point...." (In other words, giving the rats electric shocks at the point in the maze at which they had to choose, rather than before that point—the novel feature of this particular experiment—made no significant difference.) The experimenters then go on to cite work done in 1933, 1935, and other years up to 1985.[30]

The following experiment is simply an attempt to show that results already known to occur in humans also apply to mice: Curt Spanis and Larry Squire of the University of California, San Diego, used two different types of shock in one experiment designed to examine how "electroconvulsive shock" affects memory in mice. The mice were placed in the light compartment of a chamber with two compartments, the other one being dark. When the mice crossed over from the light compartment to the dark compartment their feet were given an electric shock. After "training," the mice received "electroconvulsive shock treatment...administered four times at 1-hour intervals...[and] seizures occurred in each case." The electroconvulsive shock treatment caused retrograde amnesia, which lasted at least twenty-eight days. Spanis and Squire concluded that this was the case because the mice did not remember to avoid crossing over into the dark compartment, which caused them to receive electric shocks. Spanis and Squire noted that their findings were "consistent" with findings that Squire had already made in studies based on psychiatric patients. They acknowledged that the results of the experiment "cannot strongly support or reject" ideas about memory loss because of the "high variability of the scores in the various groups." Nevertheless, they claim: "These findings extend the parallel between experimental amnesia in laboratory animals and human amnesia."[31]

In a similar experiment J. Patel and B. Migler, working at ICI Americas, Inc., in Wilmington, Delaware, trained squirrel monkeys to press a lever to obtain food pellets. The monkeys were then fitted with metal collars around their necks, through which they were given electric shocks each time they received a food pellet. They could avoid the shocks only if they waited three hours before trying to obtain food. It took eight weeks of training sessions, for six hours a day, for the monkeys to learn to avoid shocks in this way. This was supposed to produce a "conflict" situation, and the monkeys were then given various drugs to test whether monkeys on the drugs would elicit more shocks. The experimenters reported that they had also adapted the test for rats, and that it would be "useful in identifying potential anti-anxiety agents."[32]

Experiments in conditioning have been going on for over eighty-five years. A report compiled in 1982 by the New York group United Action for Animals found 1,425 papers on "classical conditioning experiments" on animals. Ironically, the futility of much of this research is grimly revealed by a paper published by a group of experimenters at the University of Wisconsin. Susan Mineka and her colleagues subjected 140 rats to shocks that could be escaped and also subjected them to shocks that could not be escaped in order to compare the levels of fear generated by such different kinds of shocks. Here is the stated rationale for their work:

> Over the past 15 years an enormous amount of research has been directed toward understanding the differential behavior and physiological effects that stem from exposure to controllable as opposed to uncontrollable aversive elements. The general conclusion has been that exposure to uncontrollable aversive events is considerably more stressful for the organism than is exposure to controllable aversive events.

After subjecting their rats to various intensities of electric shock, sometimes allowing them the possibility of escape and sometimes not, the experimenters were unable to determine what mechanisms could be considered correct in accounting for their results. Nonetheless, they said that they believed their results to be important because "they raise some question about the validity of the conclusions of the hundreds of experiments conducted over the past 15 years or so."[33]

In other words, fifteen years of giving electric shocks to animals may not have produced valid results. But in the bizarre world of psychological animal experiments, this finding serves as justification for yet more experiments giving inescapable electric shock to yet more animals so that "valid" results can finally be produced—and remember, these "valid results" will still only apply to the behavior of trapped animals subjected to inescapable electric shock.

An equally sad tale of futility is that of experiments designed to produce what is known as "learned helplessness"—supposedly a model of depression in human beings. In 1953 R. Solomon, L. Kamin, and L. Wynne, experimenters at Harvard University, placed forty dogs in a device called a "shuttlebox," which consists of a box divided into two compartments, separated by a barrier. Initially the barrier was set at the height of the dog's back. Hundreds of intense electric shocks were delivered to the dogs' feet through a grid floor. At first the dogs could escape the shock if they learned to jump the barrier into the other compartment. In an attempt to "discourage" one dog from jumping, the experimenters forced the dog to jump one hundred times onto a grid floor in the other compartment that also delivered a shock to the dog's feet. They said that as the dog jumped he gave a "sharp anticipatory yip which turned into a yelp when he landed on the electrified grid." They then blocked the passage between the compartments with a piece of plate glass and tested the dog again. The dog "jumped forward and smashed his head against the glass." The dogs began by showing symptoms such as "defecation, urination, yelping and shrieking, trembling, attacking the apparatus, and so on; but after ten or twelve days of trials dogs who were prevented from escaping shock ceased to resist. The experimenters reported themselves "impressed" by this, and concluded that a combination of the plate glass barrier and foot shock was "very effective" in eliminating jumping by dogs.[34]

This study showed that it was possible to induce a state of hopelessness and despair by repeated administration of severe inescapable shock. Such "learned helplessness" studies were further refined in the 1960s. One prominent experimenter was Martin Seligman of the University of Pennsylvania. He electrically shocked dogs through a steel grid floor with such intensity and persistence that the dogs stopped trying to escape and "learned" to be helpless. In one study, written with colleagues Steven Maier and James Geer, Seligman describes his work as follows:

> When a normal, naive dog receives escape/avoidance training in a shuttlebox, the following behavior typically occurs: at the onset of electric shock the dog runs frantically about, defecating, urinating, and howling until it scrambles over the barrier and so escapes from shock. On the next trial the dog, running and howling, crosses the barrier more quickly, and so on, until efficient avoidance emerges.

Seligman altered this pattern by strapping dogs in harnesses and giving them shocks from which they had no means of escape. When the dogs were then placed in the original shuttlebox situation from which escape was possible, he found that

> such a dog reacts initially to shock in the shuttlebox in the same manner as the naive dog. However in dramatic contrast to the naive dog it soon stops running and remains silent until shock terminates. The dog does not cross the barrier and escape from shock. Rather it seems to "give up" and passively "accept" the shock. On succeeding trials the dog continues to fail to make escape movements and thus takes 50 seconds of severe, pulsating shock on each trial.... A dog previously exposed to inescapable shock... may take unlimited shock without escaping or avoiding at all.[35]

In the 1980s, psychologists have continued to carry out these "learned helplessness" experiments. At Temple University in Philadelphia, Philip Bersh and three other experimenters trained rats to recognize a warning light that alerted them to a shock that would be delivered within five seconds. Once they understood the warning, the rats could avoid the shock by moving into the safe compartment. After the rats had learned this avoidance behavior, the experimenters walled off the safe chamber and subjected them to prolonged periods of inescapable shock. Predictably, they found that even after escape was possible, the rats were unable to relearn the escape behavior quickly.[36]

Bersh and colleagues also subjected 372 rats to aversive shock testing to try to determine the relationship between Pavlovian conditioning and learned helplessness. They reported that the "implications of these findings for learned helplessness theory are not entirely clear" and that "a substantial number of questions remain."[37]

At the University of Tennessee at Martin, G. Brown, P. Smith, and R. Peters went to a lot of trouble to create a specially designed shuttlebox for goldfish, perhaps to see if Seligman's theory holds water. The experimenters subjected forty-five fish to sixty-five shock sessions each and concluded that "the data in the present study do not provide much support for Seligman's hypothesis that helplessness is learned."[38]

These experiments have inflicted acute, prolonged pain on many animals, first to prove a theory, then to disprove the theory, and finally to support modified versions of the original theory. Steven Maier, who with Seligman and Geer was a coauthor of the previously quoted report on inducing learned helplessness in dogs, has made a career out of perpetuating the learned helplessness model. Yet in a recent review article, Maier had this to say about the validity of this "animal model" of depression:

> It can be argued that there is not enough agreement about the characteristics, neurobiology, induction, and prevention/cure of depression to make such comparison meaningful.... It would thus appear unlikely that learned helplessness is a model of depression in any general sense.[39]

Although Maier tries to salvage something from this dismaying conclusion by saying that learned helplessness may constitute a model not of depression but of "stress and coping," he has effectively admitted that more than thirty years of animal experimentation have been a waste of time and of substantial amounts of taxpayers' money, quite apart from the immense amount of acute physical pain that they have caused.

In the first edition of this book, I reported on an experiment performed at Bowling Green University in Ohio by P. Badia and two colleagues, and published in 1973. In that experiment ten rats were tested in sessions that were six hours long, during which frequent shock was "at all times unavoidable and inescapable." The rats could press either of two levers within the test chamber in order to receive warning of a coming shock. The experimenters concluded that the rats did prefer to be warned of a shock.[40] In 1984 the same experiment was still being carried out. Because someone had suggested that the previous experiment could have been "methodologically unsound," P. Badia, this time with B. Abbott of Indiana University, placed ten rats in electrified chambers, subjecting them again to six-hour shock sessions. Six rats received inescapable shock at intervals of one minute, sometimes preceded by a warning. Then they were allowed to press one of two levers to receive either shocks that were preceded by a warning signal or unsignaled shocks. The remaining four rats were used in a variation of this experiment, receiving shocks at two-minute and four-minute intervals. The experimenters found, once again, that the rats preferred shock that was signaled, even if it resulted in their receiving more shocks.[41]

Electric shock has also been used to produce aggressive behavior in animals. In one study at the University of Iowa, Richard Viken and John Knutson divided 160 rats into groups and "trained" them in a stainless steel cage with an electrified floor. Pairs of rats were given electric shocks until they learned to fight by striking out at the other rat while facing each other in an upright position or by biting. It took an average of thirty training trials before the rats learned to do this immediately on the first shock. The researchers then placed the shock-trained rats in the cage of untrained rats and recorded their behavior. After one day, all the rats were killed, shaved, and examined for wounds. The experimenters concluded that their "results were not useful in understanding the offensive or defensive nature of the shock-induced response."[42]

At Kenyon College in Ohio, J. Williams and D. Lierle performed a series of three experiments to study the effects that stress control had on defensive behavior. The first experiment was based on the assumption that uncontrollable shock enhances fear. Sixteen rats were placed in plexiglass tubes and were given inescapable electric shocks to their tails. They were then placed as intruders into an already established colony of rats and their interactions with the others were recorded. In the second experiment, twenty-four rats were able to control the shock through training. In the third experiment, thirty-two rats were exposed to inescapable shock and controllable shock. The experimenters concluded:

> Although these findings and our theoretical formulations emphasize the interrelationships among shock controllability, the predictability of shock termination, conditioned stress cues, fear, and defensive behavior, further experimentation is necessary to examine the precise nature of these complex interactions.[43]

This report, published in 1986, cited earlier experimental work in this field going back to 1948.

At the University of Kansas, a unit calling itself the Bureau of Child Research has been inflicting electric shock on a variety of animals. In one experiment, Shetland ponies were deprived of water until they were thirsty and then given a water bowl that could be electrified. Two loudspeakers were placed on either side of the ponies' heads. When noise came from the left speaker, the bowl was electrified and the ponies received an electric shock if they were drinking. They learned to stop drinking when they heard the noise from the left speaker, but not from the right. Then the speakers were moved closer together, until the ponies could no longer distinguish between them and so could not avoid shock. The researchers pointed to similar experiments on white rats, kangaroo rats, wood rats, hedgehogs, dogs, cats, monkeys, opossums, seals, dolphins, and elephants, and concluded that ponies have great difficulty in distinguishing the direction of noises as compared with other animals.[44]

It is not easy to see how this research is going to benefit children. Indeed, in general, what is so disturbing about the examples of research given above is that despite the suffering the animals have gone through, the results obtained, even as reported by the experimenters themselves, are trivial, obvious, or meaningless. The conclusions of the experiments cited above show, clearly enough, that experimental psychologists have put a lot of effort into telling us in scientific jargon what we knew all along, and what we could have found out in less harmful ways with a little thought—and these experiments were supposedly more significant than others that did not get published.

We have looked at only a very small number of psychology experiments that involve electric shock. According to the Office of Technology Assessment report,

> A survey of the 608 articles appearing from 1979 through 1983 in the American Psychological Association journals that typically publish animal research identified 10 percent of the studies as using electric shock.[45]

Many other journals not associated with the American Psychological Association also publish reports of animal studies that have used electric shock; and we must not forget the experiments that never get published at all. And this is only one kind of painful or distressing research carried out on animals within the field of psychology. We have already looked at maternal deprivation studies; but one could fill several books with brief descriptions of yet more kinds of psychological experimentation, such as abnormal behavior, animal models of schizophrenia, animal movements, body maintenance, cognition, communication, predator-prey relations, motivation and emotion, sensation and perception, and sleep, food, and water deprivation. We have considered but a few of the tens of thousands of experiments performed annually in the field of psychology, but they should be enough to show that many, many experiments still being conducted cause great pain to animals and offer no prospect of yielding really momentous or vital new knowledge. Unfortunately, animals have become, for the psychologist and for other experimenters, mere tools. A laboratory may consider the cost of these "tools," but a certain callousness toward them becomes apparent, not only in the experiments performed but also in the wording of the reports. Consider, for instance, Harlow and Suomi's mention of their "rape rack" and the jocular tone in which they report on the "favorite tricks" of the female monkeys born as a result of its use.

Detachment is made easier by the use of technical jargon that disguises the real nature of what is going on. Psychologists, under the influence of the behaviorist doctrine that only what can be observed should be mentioned, have developed a considerable collection of terms that refer to pain without appearing to do so. Alice Heim, one of the few psychologists who has spoken out against the pointless animal experimentation of her colleagues, describes it this way:

> The work on "animal behavior" is always expressed in scientific, hygienic-sounding terminology, which enables the indoctrination of the normal, non-sadistic young psychology student to proceed without his anxiety being aroused. Thus techniques of "extinction" are used for what is in fact torturing by thirst or near starvation or electric-shocking; "partial reinforcement" is the term for frustrating an animal by only occasionally fulfilling the expectations which the experimenter has aroused in the animal by previous training; "negative stimulus" is the term used for subjecting an animal to a stimulus which he avoids, if possible. The term "avoidance" is O.K. because it is an observable activity. The term "painful" or "frightening" stimulus are less O.K. since they are anthropomorphic, they imply that the animal has feelings—and that these may be similar to human feelings. This is not allowable because it is non-behavioristic and unscientific (and also because this might deter the younger and less hard-boiled researcher from pursuing certain ingenious experiments. He might allow a little play to his imagination). The cardinal sin for the experimental psychologist working in the field of "animal behavior" is anthropomorphism. Yet if he did not believe in the analogue of the human being and the lower animal even he, presumably, would find his work largely unjustified.[46]

We can see the kind of jargon to which Heim refers in the reports of experiments I have already cited. Note that even when Seligman feels compelled to say that the subjects of his experiments "gave up" trying to escape shock, he finds it necessary to place the term in quotation marks, as if to say that he is not really imputing any kind of mental processes to the dog. Yet the logical consequence of this view of "scientific method" is that experiments on animals cannot teach us anything about human beings.

Amazing as it may seem, some psychologists have been so concerned to avoid anthropomorphism that they have accepted this conclusion. This attitude is illustrated by the following autobiographical statement, which appeared in *New Scientist*:

> When fifteen years ago I applied to do a degree course in psychology, a steely-eyed interviewer, himself a psychologist, questioned me closely on my motives and asked me what I believed psychology to be and what was its principal subject matter? Poor naive simpleton that I was, I replied that it was the study of the mind and that human beings were its raw material. With a glad cry at being able to deflate me so effectively, the interviewer declared that psychologists were not interested in the mind, that rats were the golden focus of study, not people, and then he advised me strongly to trot around to the philosophy department next door. . . .[47]

Perhaps not many psychologists would now proudly state that their work has nothing to do with the human mind. Nevertheless many of the experiments that are performed on rats can only be explained by assuming that the experimenters really are interested in the behavior of the rat for its own sake, without any thought of learning anything about humans. In that case, though, what possible justification can there be for the infliction of so much suffering? It is certainly not for the benefit of the rat.

So the researcher's central dilemma exists in an especially acute form in psychology: either the animal is not like us, in which case there is no reason for performing the experiment; or else the animal is like us, in which case we ought not to perform on the animal an experiment that would be considered outrageous if performed on one of us.

Another major field of experimentation involves the poisoning of millions of animals annually. Often this too is done for trivial reasons. In Britain in 1988, 588,997 scientific procedures were performed on animals to test drugs and other materials; of these, 281,358 were not related to the testing of medical or veterinary products.[48] In the United States no accurate figures are available, but if the proportion is similar to Britain the number of animals used in testing must be at least three million. In fact it is probably double or triple that figure, because there is so much research and development in this field in the United States and the Food and Drug Administration requires extensive testing of new substances before they are released. It may be thought justifiable to require tests on animals of potentially life-saving drugs, but the same kinds of tests are used for products like cosmetics, food coloring, and floor polishes. Should thousands of animals suffer so that a new kind of lipstick or floor wax can be put on the market? Don't we already have an excess of most of these products? Who benefits from their introduction, except the companies that hope to profit from them?

In fact, even when the test is carried out on a medical product, it is most probably not going to do anything to improve our health. Scientists working for the British Department of Health and Social Security examined drugs marketed in Britain between 1971 and 1981. New drugs, they found,

> have largely been introduced into therapeutic areas already heavily oversubscribed ... for conditions which are common, largely chronic and occur principally in the affluent Western Society. Innovation is therefore largely directed towards commercial returns rather than therapeutic need.[49]

To appreciate what is involved in introducing all these new products it is necessary to know something about the standard methods of testing. In order to determine how poisonous a substance is, "acute oral toxicity tests" are performed. These tests, developed in the 1920s, force animals to ingest substances, including nonedible products such as lipstick and paper. Often the animals will not eat the substance if it is simply placed in their food, so experimenters either force-feed the animals by mouth or insert a tube down their throats. Standard tests are carried out for fourteen days but some may last for up to six months—if the animals survive that long. During this time, the animals often display classic symptoms of poisoning, including vomiting, diarrhea, paralysis, convulsions, and internal bleeding.

The most widely known acute toxicity test is the LD50. LD50 stands for "lethal dose 50 percent": the amount of the substance that will kill half of the animals in the study. To find that dose level, sample groups of animals are poisoned. Normally, before the point at which half of them die is reached, the animals are all very ill and in obvious distress. In the case of fairly harmless substances it is still considered good procedure to find the concentration that will make half the animals die; consequently enormous quantities have to be force-fed to the animals, and death may be caused merely by the large volume or high concentration given to the animals. This has no relevance to the circumstances in which humans will use the product. Since the very point of these experiments is to measure how much of the substance will poison half the animals to death, dying animals are not put out of their misery for fear of producing inaccurate results. The U.S. Congress Office of Technology Assessment has estimated that "several million" animals are used each year for toxicological testing in the United States. No more specific estimates for the LD50 test are available.[50]

Cosmetics and other substances are tested in animals' eyes. The Draize eye irritancy tests were first used in the 1940s, when J. H. Draize, working for the U.S. Food and Drug Administration, developed a scale for assessing how irritating a substance is when placed in rabbits' eyes. The animals are usually placed in holding devices from which only their heads protrude. This prevents them scratching or rubbing their eyes. A test substance (such as bleach, shampoo, or ink) is then placed in one eye of each rabbit. The method used is to pull out the lower eyelid and place the substance into the small "cup" thus formed. The eye is then held closed. Sometimes the application is repeated. The rabbits are observed daily for eye swelling, ulceration, infection, and bleeding. The studies can last up to three weeks. One researcher employed by a large chemical company has described the highest level of reaction as follows:

> Total loss of vision due to serious internal injury to cornea or internal structure. Animal holds eye shut urgently. May squeal, claw at eye, jump and try to escape.[51]

But, of course, when in the holding device the rabbits can neither claw at their eyes nor escape (see photograph following page 157). Some substances cause such serious damage that the rabbits' eyes lose all distinguishing characteristics—the iris, pupil, and cornea begin to resemble one massive infection. Experimenters are not obliged to use anesthetics, but sometimes they will use a small amount of topical anesthetic when introducing the substance, provided it does not interfere with the test. This does nothing to alleviate the pain that can result after two weeks of having oven cleaner in the eye. U.S. Department of Agriculture figures show that, in 1983, toxicology testing laboratories used 55,785 rabbits, and chemical companies an additional 22,034. It can be assumed that many of these were used for Draize tests, although no estimate of the number is available.[52]

Animals are also subjected to other tests to determine the toxicity of many substances. During inhalation studies, animals are placed in sealed chambers and forced to inhale sprays, gases, and vapors. In dermal toxicity studies, rabbits have their fur removed so that a test substance can be placed on their skin. The animals are restrained so that they do not scratch at their irritated bodies. The skin may bleed, blister, and peel. Immersion studies, in which animals are placed in vats of diluted substances, sometimes cause the animals to drown before any test results can be obtained. In injection studies, the test substance is injected directly into the animal, either under the skin, into the muscles, or directly into an organ.

These are the standard procedures. Here are two examples of how they are carried out:

In England, the Huntingdon Research Institute, together with the giant corporation ICI, carried out experiments in which forty monkeys were poisoned with the weed-killer paraquat. They became very ill, vomited, had difficulty in breathing, and suffered from hypothermia. They died slowly, over several days. It was already known that paraquat poisoning in humans results in a slow and agonizing death.[53]

We began this chapter with some military experiments. Here is a military experiment involving an LD50 test:

Experimenters at the U.S. Army Medical Research Institute of Infectious Diseases poisoned rats with T-2. This is a poison which, according to the Department of State, has "the added advantage of being an effective terror weapon that causes bizarre and horrifying symptoms" such as "severe bleeding," blisters, and vomiting, so that humans and animals may be "killed in a gruesome manner." The T-2 was administered intramuscularly, intravenously, subcutaneously, interperitoneally—i.e., injected into the muscle tissue, into the veins, under the skin, and into the lining of the abdomen—and through the nose and mouth, and on the skin. All eight tests were to determine LD50 values. Death usually occurred between nine and eighteen hours after exposure, but the rats exposed through the skin took an average of six days to die. Before death the animals were unable to walk or eat, had rotting of the skin and intestines, restlessness, and diarrhea. The experimenters reported that their findings were "quite compatible with earlier published studies of subacute and chronic exposure to T-2."[54]

As this example illustrates, it is not only products intended for human consumption that are tested. Chemical warfare agents, pesticides, and all kinds of industrial and household goods are fed to animals or put in their eyes. A reference book, *Clinical Toxicology of Commercial Products*, provides data, mostly from animal experiments, on how poisonous hundreds of commercial products are. The products include insecticides, antifreeze, brake fluid, bleaches, Christmas tree sprays, church candles, oven cleaners, deodorants, skin fresheners, bubble baths, depilatories, eye makeup, fire extinguishers, inks, suntan oils, nail polish, mascara, hair sprays, paints, and zipper lubricants.[55]

Many scientists and physicians have criticized this type of testing, pointing out that the results are inapplicable to human beings. Dr. Christopher Smith, a physician from Long Beach, California, has said:

> The results of these tests cannot be used to predict toxicity or to guide therapy in human exposure. As a board-certified emergency medicine physician with over 17 years of experience in the treatment of accidental poisoning and toxic exposures, I know of no instance in which an emergency physician has used Draize test data to aid in the management of an eye injury. I have never used results from animal tests to manage accidental poisoning. Emergency physicians rely on case reports, clinical experience and experimental data from clinical trials in humans when determining the optimal course of treatment for their patients.[56]

Toxicologists have known for a long time that extrapolation from one species to another is a highly risky venture. The most notorious drug to have caused unexpected harm to humans is thalidomide—which was extensively tested on animals before it was released. Even after thalidomide was suspected of causing deformities in humans, laboratory tests on pregnant dogs, cats, rats, monkeys, hamsters, and chickens all failed to produce deformities. Only when a particular strain of rabbit was tried were deformities produced.[57] More recently, Opren passed all the usual animal tests before it was released and extensively touted as a new "wonder drug" for the treatment of arthritis by its manufacturer, the pharmaceutical giant Eli Lilly. Opren was suspended from use in Britain after sixty-one deaths and over 3,500 reports of adverse reactions. A report in *New Scientist* estimated that the real toll could have been much higher.[58] Other drugs that were considered safe after animal tests but later proved harmful are the heart disease drug Practolol, which caused blindness, and the cough suppressant Zipeprol, which produced seizures and comas in some of those who took it.[59]

As well as exposing people to harm, testing on animals may lead us to miss out on valuable products that are dangerous to animals but not to human beings. Insulin can produce deformities in infant rabbits and mice, but not in humans.[60] Morphine, which is calming to human beings, causes mice to go into drug frenzies. And as another toxicologist has said: "If penicillin had been judged by its toxicity on guinea pigs, it might never have been used on man."[61]

After decades of mindless animal testing, there are now some signs of second thoughts. As Dr. Elizabeth Whelan, a scientist and executive director of the American Council on Science and Health, has pointed out: "It doesn't take a Ph.D. in the sciences to grasp the fact that rodent exposure to the saccharin equivalent of 1,800 bottles of soda pop a day doesn't relate well to our daily ingestion of a few glasses of the stuff." Whelan has welcomed the fact that officials at the Environmental Protection Agency recently downgraded earlier estimates of risks of pesticides and other environmental chemicals, noting that the evaluation of cancer risk, derived from animal extrapolation, was based on "simplistic" assumptions that "strain credibility." This means, she says, that "our regulators are beginning to take note of the scientific literature rejecting the infallibility of the laboratory animal test."[62]

The American Medical Association has also admitted that animal models have questionable accuracy. An AMA representative testified at a congressional hearing on drug testing that "frequently animal studies prove little or nothing and are very difficult to correlate to humans."[63]

Fortunately much progress has been made in eliminating such animal testing since the first edition of this book appeared. Most scientists then did not take seriously the possibility that effective substitutes could be found for tests that use animals to measure toxicity. They were persuaded to do so by the hard work of a large number of opponents of animal experiments. Prominent among them was Henry Spira, a former civil rights activist who put together coalitions against the Draize and LD50 tests. The Coalition to Abolish the Draize Test began by inviting Revlon, as the largest cosmetics company in the United States, to put one tenth of one percent of its profits toward developing an alternative to the Draize test. When Revlon declined, full-page advertisements appeared in *The New York Times* asking "HOW MANY RABBITS DOES REVLON BLIND FOR BEAUTY'S SAKE?"[64] People in rabbit costumes appeared at Revlon's annual general meeting. Revlon got the message and allocated the requested funds to pay for research on alternatives to animal experiments. Other companies, such as Avon and Bristol-Myers, followed suit.[65] As a result, early British work in this field by the Fund for the Replacement of Animals in Medical Experiments was taken up on a larger scale in the United States, especially at the Johns Hopkins Center for Alternatives to Animal Testing, in Baltimore. Increasing interest led to the launching of several major new journals, such as *In-Vitro Toxicology, Cell Biology and Toxicology,* and *Toxicology in Vitro.*

It took some time for this work to show results, but gradually the interest in alternatives grew. Corporations such as Avon, Bristol-Myers, Mobil, and Procter & Gamble began using alternatives in their own laboratories, thus reducing the number of animals used. Toward the end of 1988, the rate of change began to quicken. In November, an international campaign against Benetton led by the Washington, D.C., organization People for the Ethical Treatment of Animals persuaded the fashion chain to stop using animal tests in its cosmetics division.[66] In December 1988 Noxell Corporation, manufacturer of Noxzema skin creams and Cover Girl cosmetics, announced that it would use a screening test that would reduce by 80 to 90 percent the number of animals otherwise used for eye safety testing; subsequently Noxell stated that it used no animals at all in safety tests during the first half of 1989.[67]

Now the momentum was building fast. In April 1989, Avon announced that it had validated tests using a specially developed synthetic material called Eytex as a replacement for the Draize test. As a result, nine years after Spira began his campaign Avon ceased to use the Draize test.[68] Still more good news was to come. In May 1989 both Mary Kay Cosmetics and Amway announced that they had ceased to use laboratory animals for consumer product safety testing while they reviewed plans for using alternatives.[69] In June, Avon, under pressure from another campaign led by People for the Ethical Treatment of Animals, announced a permanent end to all animal testing.[70] Eight days after the Avon announcement, Revlon said that it had completed its long-term plan to eliminate animal testing in all phases of research, development, and manufacturing of all its products, and therefore it was ending animal testing. Then Fabergé abandoned the use of animals for testing in its cosmetics and toiletries business. Thus within a few months (though on the basis of many years of work) the first, second, and fourth largest United States cosmetics companies had given up all animal testing.[71]

Although the most dramatic developments have taken place in the highly public and therefore relatively vulnerable cosmetics industry, the movement against animal testing is also taking effect in wider areas of industry. As a report in *Science* put it:

> Prodded by the animal welfare movement, major manufacturers of pharmaceuticals, pesticides and household products have made significant advances in recent years toward the goal of reducing the number of animals used in toxicity testing. Alternative methods, such as cell and tissue culture and computer modeling, are increasingly being seen not just

as good public relations but as desirable both economically and scientifically.[72]

The report went on to quote Gary Flamm, director of the Food and Drug Administration Office of Toxicology Sciences, as saying that the LD50 "should be replaceable in the vast majority of cases." A *New York Times* article quoted a senior toxicologist at G. D. Searle and Company as admitting that "an awful lot of the points made by the animal welfare movement are extreme but right."[73]

There seems to be little doubt that as a result of all these developments, an immense amount of needless pain and suffering has been avoided. Precisely how much is hard to say, but millions of animals would have suffered each year in tests that will now not be performed. The tragedy is that if only the toxicologists, the corporations, and the regulatory agencies had cared more about the animals they were using, millions of animals could have been spared acute pain. It was not until the Animal Liberation movement began to make people aware of the issue that those in charge of the testing business really thought about animal suffering. The most callous, stupid things were done just because regulations required them; and no one bothered to try to change the regulations. It was not until 1983, for example, that U.S. federal agencies stated that substances known to be caustic irritants, such as lye, ammonia, and oven cleaners, did not need to be tested on the eyes of conscious rabbits.[75] But the battle is by no means over. To quote once more from the report in *Science* of April 17, 1987:

> Unnecessary testing is still wasting a lot of animals, not only because of outmoded requirements but because much existing information is not easily accessible. Theodore M. Farber, director of the [U.S. Environmental Protection Agency's] Toxicology Branch said that his agency has files of 42,000 completed tests, and 16,000 LD50 tests. He said these could be of far more use in eliminating redundant tests if they were computerized for easy accessibility. "Many of us in regulatory toxicology see the same studies over and over again," said Farber.

Stopping this waste of animal lives and animal pain should not be difficult, if people really want to do it. Developing completely adequate alternatives to all tests for toxicity will take longer, but it should be possible. Meanwhile there is a simple way to cut down the amount of suffering involved in such tests. Until we have developed satisfactory alternatives, as a first step we should just do without any new but potentially hazardous substances that are not essential to our lives.

When experiments can be brought under the heading "medical" we are inclined to think that any suffering they involve must be justifiable because the research is contributing to the alleviation of suffering. But we have already seen that the testing of therapeutic drugs is less likely to be motivated by the desire for maximum good to all than by the desire for maximum profit. The broad label "medical research" can also be used to cover research that is motivated by a general intellectual curiosity. Such curiosity may be acceptable as part of a basic search for knowledge when it involves no suffering, but should not be tolerated if it causes pain. Very often, too, basic medical research has been going on for decades and much of it, in the long run, turns out to have been quite pointless. As an illustration, consider the following series of experiments stretching back nearly a century, on the effects of heat on animals:

In 1880 H. C. Wood placed a number of animals in boxes with glass lids and placed the boxes on a brick pavement on a hot day. He used rabbits, pigeons, and cats. His observations on a rabbit are typical. At a temperature of 109.5 degrees Fahrenheit the rabbit jumps and "kicks hind legs with great fury." The rabbit then has a convulsive attack. At 112 degrees Fahrenheit the animal lies on its side slobbering. At 120 degrees Fahrenheit it is gasping and squealing weakly. Soon after it dies.[76]

In 1881 a report appeared in *The Lancet* on dogs and rabbits whose temperatures had been raised to 113 degrees Fahrenheit. It was found that death could be prevented by cool air currents, and the results were said to indicate "the importance of keeping down the temperature in those cases in which it exhibits a tendency to rise to [an] extreme height."[77]

In 1927 W. W. Hall and E. G. Wakefield of the U.S. Naval Medical School placed ten dogs in a hot humid chamber to produce experimental heatstroke. The animals first showed restlessness, breathing difficulties, swelling and congestion of the eyes, and thirst. Some had convulsions. Some died early in the experiment. Those who did not had severe diarrhea and died after removal from the chamber.[78]

In 1954 at Yale University School of Medicine, M. Lennox, W. Sibley, and H. Zimmerman placed thirty-two kittens in a "radiant-heating" chamber. The kittens were "subjected to a total of 49 heating periods.... Struggling was common, particularly as the temperature rose." Convulsions occurred on nine occasions: "Repeated convulsions were the rule." As many as thirty convulsions occurred in rapid sequence. Five kittens died during convulsions, and six without convulsions. The other kittens were killed by the experimenters for autopsies. The experimenters reported: "The findings in artificially induced fever in kittens conform to the clinical and EEG findings in human beings and previous clinical findings in kittens."[79]

The following experiment was performed at the K. G. Medical College, Lucknow, India. I include it as an example of the triumph of Western methods of research and attitudes to animals over the ancient tradition of Hinduism, which has more respect for nonhuman animals than the Judeo-Christian tradition. In 1968 K. Wahal, A. Kumar, and P. Nath exposed forty-six rats to high temperature for four hours. The rats became restless, breathed with difficulty, and salivated profusely. One animal died during the experiment and the others were killed by the experimenters because "they could not survive anyway."[80]

In 1969 S. Michaelson, a veterinarian at the University of Rochester, exposed dogs and rabbits to heat-producing microwaves until their temperatures reached the critical level of 107 degrees Fahrenheit or greater. He observed that dogs start panting shortly after microwave exposure begins. Most "display increased activity varying from restlessness to extreme agitation." Near the point of death, weakness and prostration occur. In the case of rabbits "within 5 minutes, desperate attempts are made to escape the cage," and the rabbits die within forty minutes. Michaelson concluded that an increase in heat from microwaves produces damage "indistinguishable from fever in general."[81]

At the Heller Institute of Medical Research, Tel Aviv, Israel, in experiments published in 1971 and paid for by the United States Public Health Service, T. Rosenthal, Y. Shapiro, and others placed thirty-three dogs "randomly procured from the local dog pound" in a temperature-controlled chamber and forced them to exercise on a treadmill in temperatures as high as 113 degrees Fahrenheit until "they collapsed in heatstroke or reached a predetermined rectal temperature." Twenty-five of the dogs died. Nine more dogs were then subjected to a temperature of 122 degrees Fahrenheit without treadmill exercise. Only two of these dogs survived longer than twenty-four hours, and autopsies showed that all had hemorrhaged. The experimenters concluded: "The findings are in accordance with what is reported in the literature on humans."[82] In a further report published in 1973, the same researchers describe experiments on fifty-three dogs, involving various combinations of heat and treadmill exercise. Six of the dogs vomited, eight had diarrhea, four went into convulsions, twelve lost muscle coordination, and all salivated excessively. Of ten dogs whose rectal temperature reached 113 degrees Fahrenheit, five died "at the moment of maximum rectal temperature" and the other five died between thirty minutes and eleven hours after the end of the experiment. The experimenters concluded that "the sooner the heatstroke victim's temperature is brought down, the greater the chances of recovery."[83]

In 1984 experimenters working for the Federal Aviation Administration, stating that "animals occasionally die from heat stress encountered during shipping in the nation's transportation systems," subjected ten beagles to experimental heat. The dogs were isolated in chambers, fitted with muzzles, and exposed to 95 degrees Fahrenheit combined with high humidity. They were given no food or water, and were kept in these conditions for twenty-four hours. The behavior of the dogs was observed; it included "deliberate agitated activity such as pawing at the crate walls, continuous circling, tossing of the head to shed the muzzle, rubbing the muzzle back and forth on the floor of the crate, and aggressive acts on the sensor guards." Some of the dogs died in the chambers. When the survivors were removed, some vomited blood, and all were weak and exhausted. The experimenters refer to "subsequent experiments on more than 100 beagles."[84]

In a further example of military experimentation, R. W. Hubbard, of the U.S. Army Research Institute of Environmental Medicine in Natick, Massachusetts, has been publishing papers with titles such as "Rat Model of Acute Heatstroke Mortality" for more than a decade. It is well known that when rats are hot they spread saliva over their bodies; the saliva plays the same cooling role as sweating in humans. In 1982 Hubbard and two colleagues noted that rats unable to produce saliva will spread urine if no other fluid is available.[85] So in 1985 the same three researchers, joined by a fourth, injected rats with the drug atropine, which inhibits both sweating and the secretion of saliva. Other rats had their salivary glands removed by surgery. The experimenters then placed the rats in chambers at 107 degrees Fahrenheit until their body temperature rose to 108.7 degrees Fahrenheit. The researchers drew diagrams comparing the "urine spreading pattern" of a rat who had either been given atropine or had been surgically "desalivated" with that of an untreated rat. They found the "atropinized heat-stressed rat model" to be "a promising tool with which to examine the role of dehydration in heat illness."[86]

Here we have cited a series of experiments going back into the nineteenth century—and I have had space sufficient to include only a fraction of the published literature. The experiments obviously caused great suffering; and the major finding seems to be the advice that heatstroke victims should be cooled—something that seems to be fairly elementary common sense and in any case had already been borne out by observations on human beings who have suffered natural heatstroke. As for the application of this research to human beings, B. W. Zweifach showed in 1961 that dogs are physiologically different from human beings in ways that affect their response to heatstroke, and hence they are a poor model for heatstroke in human beings.[87] It is hard to take seriously the suggestion that small furry animals drugged with atropine who spread urine over themselves when hot will be a better model.

Similar series of experiments are to be found in many other fields of medicine. In the New York City offices of United Action for Animals there are filing cabinets full of photocopies of experiments reported in the journals. Each thick file contains reports on numerous experiments, often fifty or more, and the labels on the files tell their own story: "Acceleration," "Aggression," "Asphyxiation," "Blinding," "Burning," "Centrifuge," "Compression," "Concussion," "Crowding," "Crushing," "Decompression," "Drug Tests," "Experimental Neurosis," "Freezing," "Heating," "Hemorrhage," "Hindleg Beating," "Immobilization," "Isolation," "Multiple Injuries," "Prey Killing," "Protein Deprivation," "Punishment," "Radiation," "Starvation," "Shock," "Spinal Cord Injuries," "Stress," "Thirst," and many more. While some of the experiments may have led to advances in medical knowledge, the value of this knowledge is often questionable, and in some cases the knowledge might have been gained in other ways. Many of the experiments appear to be trivial or misconceived, and some of them were not even designed to yield important benefits.

Consider, as another example of the way in which endless variations of the same or similar experiments are carried out, these experiments relating to the experimental production of shock in animals (by which is meant not electric shock but the mental and physical state of shock that often occurs after a severe injury). As long ago as 1946 a researcher in the field, Magnus Gregersen of Columbia University, surveyed the literature and found over eight hundred published papers dealing with experimental studies of shock. He describes the methods used to induce shock:

> The use of a tourniquet on one or more extremities, crush, compression, muscle trauma by contusion with light hammer blows, Noble-Collip drum [a device in which animals are placed and the drum rotated; the animals tumble repeatedly to the bottom of the drum and injure themselves], gunshot wounds, strangulation or intestinal loops, freezing, and burns.

Gregersen also notes that hemorrhage has been "widely employed" and "an increasing number of these studies has been done without the complicating factor of anesthesia." He is not, however, pleased by all this diversity, and complains that the variety of methods makes it "exceedingly difficult" to evaluate the results of different researchers; there is, he says, a "crying need" for standardized procedures that will invariably produce a state of shock.[88]

Eight years later the situation had not changed much. S. M. Rosenthal and R. C. Millican wrote that "animal investigations in the field of traumatic shock have yielded diversified and often contradictory results." Nevertheless they looked forward to "future experimentation in this field" and like Gregersen they discouraged the use of anesthesia: "The influence of anesthesia is controversial... [and] in the reviewers' opinion prolonged anesthesia is best avoided...." They also recommended that "adequate numbers of animals must be employed to overcome biological variations."[89]

In 1974 experimenters were still working on "animal models" of experimental shock, still carrying out preliminary experiments to determine what injuries might be inflicted to produce a satisfactory "standard" state of shock. After decades of experiments designed to produce shock in dogs by causing them to hemorrhage, more recent studies indicated that (surprise!) hemorrhage-induced shock in dogs is not like shock in humans. Noting these studies, researchers at the University of Rochester caused hemorrhage in pigs, which they think may be more like humans in this respect, to determine what volume of blood loss might be suitable for the production of experimental shock.[90]

Hundreds of experiments are also performed annually in which animals are forced to become addicted to drugs. On cocaine alone, for example, over 500 studies have been conducted. An analysis of just 380 of these estimated that they cost about $100 million, most of it tax money.[91] Here is one example:

In a laboratory at Downstate Medical Center run by Gerald Deneau, rhesus monkeys were locked into restraining chairs. The animals were then taught to self-administer cocaine directly into the bloodstream in whatever quantities they wanted by pushing a button. According to one report,

> the test monkeys pushed the button over and over, even after convulsions. They went without sleep. They ate five to six times their normal amount, yet became emaciated.... In the end, they began to mutilate themselves and, eventually, died of cocaine abuse.

Dr. Deneau has acknowledged that "few people could afford the massive doses of cocaine these monkeys were able to obtain."[92]

Even though five hundred animal experiments have been conducted involving cocaine, this is only a small part of the total amount of experimentation that involves turning animals into addicts. In the first edition of this book I reported on a similar set of addiction experiments, using morphine and amphetamines. Here are some more recent examples:

At the University of Kentucky, beagles were used to observe withdrawal symptoms from Valium and a similar tranquillizer called Lorazepam. The dogs were forced to become addicted to the drug and then, every two weeks, the tranquillizers were withdrawn. Withdrawal symptoms included twitches, jerks, gross body tremors, running fits, rapid weight loss, fear, and cowering. After forty hours of Valium withdrawal, "numerous tonic-clonic convulsions were seen in seven of nine dogs.... Two dogs had repeated episodes of clonic seizures involving the whole body." Four of the dogs died—two while convulsing and two after rapid weight loss. Lorazepam produced similar symptoms but not convulsive deaths. The experimenters reviewed experiments going back to 1931 in which barbiturate and tranquillizer withdrawal symptoms had been observed in rats, cats, dogs, and primates.[93]

After reviewing the history of experiments showing that "withdrawal-like effects can occur following single administrations of opiates in several species," including dogs, mice, monkeys, and rats, D. M. Grilly and G. C. Gowans of Cleveland State University proceeded to test a hypothesis that morphine withdrawal produces hypersensitivity to pain. Rats were trained by a procedure that involved an average of 6,387 training trials in "shock discrimination." In these trials, the rats had to respond to receiving an electric shock. The rats were then injected with morphine and exposed to electric shocks one, two, three, and seven days after. The experimenters noted that sensitivity to shock was elevated during the days immediately following morphine administration.[94]

Here is an even more bizarre example of drug research:

At the University of California at Los Angeles, Ronald Siegel chained two elephants to a barn. The female elephant was used in range-finding tests "to determine procedures and dosages for LSD administration." She was given the drug orally and by dart-gun. After this the experimenters dosed both elephants every day for two months and observed their behavior. High doses of the hallucinogen caused the female to fall down on her side, trembling and barely breathing, for one hour. The high doses caused the bull elephant to become aggressive and charge Siegel, who described such repeated aggressive behavior as "inappropriate."

My final episode in this grim tale of drug experimentation does, at least, have a happy ending. Researchers at Cornell University Medical College fed large doses of barbiturates to cats by means of tubes surgically implanted in their stomachs. They then abruptly stopped the barbiturates. Here is their description of the withdrawal symptoms:

> Some were unable to stand.... The "spread eagle posture" was seen in animals displaying the most severe abstinence signs and the most frequent grand mal type convulsions. Almost all of these animals died during or soon after periods of continuous convulsive activity.... Rapid or labored respiration was often noted when other abstinence signs were most intense.... Hypothermia was noted when animals were weakest, especially after persistent seizures and when near death.[96]

These experiments began in 1975. Although barbiturate abuse had been a serious problem a few years earlier, by that time the use of barbiturates was severely restricted, and abuse had declined. It has continued to do so since. Nevertheless, the cat experiments at Cornell continued for fourteen years. Then, in 1987, Trans-Species Unlimited, a Pennsylvania-based animal rights group, compiled all the available information they could find about the experiments and began to campaign to stop them. For four months, concerned people picketed the laboratory at which the cat studies were being conducted and wrote letters to the funding agencies, the press, the university, and legislators. After defending the experiments for a long time, late in 1988 Cornell and Michiko Okamoto, the researcher carrying out the experiments, wrote to the funding body, the National Institute on Drug Abuse, to say they would forfeit a new $530,000 research grant that would have paid for three more years of experiments.[97]

How can these things happen? How can people who are not sadists spend their working days driving monkeys into lifelong depression, heating dogs to death, or turning cats into drug addicts? How can they then remove their white coats, wash their hands, and go home to dinner with their families? How can taxpayers allow their money to be used to support these experiments? How did students carry on protests against injustice, discrimination, and oppression of all kinds, no matter how far from home, while ignoring the cruelties that were—and still are—being carried out on their own campuses?

The answer to these questions lies in the unquestioned acceptance of speciesism. We tolerate cruelties inflicted on members of other species that would outrage us if performed on members of our own species. Speciesism allows researchers to regard the animals they experiment on as items of equipment, laboratory tools rather than living, suffering creatures. In fact, on grant applications to government funding agencies, animals are listed as "supplies" alongside test tubes and recording instruments.

In addition to the general attitude of speciesism that experimenters share with other citizens, some special factors also help to make possible the experiments I have described. Foremost among these is the immense respect that people still have for scientists. Although the advent of nuclear weapons and environmental pollution has made us realize that science and technology are not as beneficial as they might appear at first glance, most people still tend to be in awe of anyone who wears a white coat and has a Ph.D. In a well-known series of experiments Stanley Milgram, a Harvard psychologist, demonstrated that ordinary people will obey the directions of a white-coated researcher to administer what appears to be (but in fact is not) electric shock to a human subject as "punishment" for failing to answer questions correctly, and they will continue to do this even when the human subject cries out and pretends to be in great pain.[98] If this can happen when the participants believe they are inflicting pain on a human being, how much easier is it for students to push aside their initial qualms when their professors instruct them to perform experiments on animals? What Alice Heim has rightly called the "indoctrination" of the student is a gradual process, beginning with the dissection of frogs in school biology classes. When the future medical students, psychology students, or veterinarians reach the university and find that to complete the course of studies on which they have set their hearts they must experiment on living animals, it is difficult for them to refuse to do so, especially since they know that what they are being asked to do is standard practice. Those students who have refused to engage in such studies have found themselves failing their courses and are often forced to leave their chosen field of study.

The pressure to conform does not let up when students receive their degrees. If they go on to graduate degrees in fields in which experiments on animals are usual, they will be encouraged to devise their own experiments and write them up for their Ph.D. dissertations. Naturally, if this is how students are educated they will tend to continue in the same manner when they become professors, and they will, in turn, train their own students in the same manner.

Here the testimony of Roger Ulrich, a former experimenter who escaped from his conditioning and now acknowledges that he inflicted "years of torture" on animals from rats to monkeys, is particularly revealing. In 1977 the magazine *Monitor*, published by the American Psychological Association, reported that experiments on aggression carried out by Ulrich had been singled out before a congressional subcommittee as an example of inhumane research. To the surprise of the antivivisectionists who had criticized him, and no doubt to the editor of the *Monitor* as well, Ulrich wrote back to say that he was "heartened" by the criticism, and added:

> Initially my research was prompted by the desire to understand and help solve the problem of human aggression, but I later discovered that the results of my work did not seem to justify its continuance. Instead I began to wonder if perhaps financial rewards, professional prestige, the opportunity to travel, etc. were the maintaining factors, and if we of the scientific community (supported by our bureaucratic and legislative system) were actually a part of the problem.[99]

Don Barnes, who as we saw had a similar change of mind about his work irradiating trained monkeys for the U.S. Air Force, calls the process Ulrich describes "conditioned ethical blindness." In other words, just as a rat can be conditioned to press a lever in return for a reward of food, so a human being can be conditioned by professional rewards to ignore the ethical issues raised by animal experiments. As Barnes says:

> I represented a classic example of what I choose to call "conditioned ethical blindness." My entire life had consisted of being rewarded for using animals, treating them as sources of human improvement or amusement.... During my sixteen years in the laboratory the morality and ethics of using laboratory animals were never broached in either formal or informal meetings prior to my raising the issues during the waning days of my tenure as a vivisector.[100]

It is not only the experimenters themselves who suffer from conditioned ethical blindness. Research institutions sometimes answer critics by telling them that they employ a veterinarian to look after the animals. Such statements are supposed to provide reassurance, because of the widespread belief that all veterinarians are people who care about animals and would never let them suffer unnecessarily. Regrettably, this is not the case. No doubt many veterinarians did go into the field because they cared about animals, but it is difficult for people who really care about animals to go through a course of study in veterinary medicine without having their sensitivity to animal suffering blunted. Those who care most may not be able to complete their studies. One former veterinary student wrote to an animal welfare organization:

> My life-long dream and ambition to become a veterinarian dissipated following several traumatic experiences involving standard experimental procedures utilized by the dispassionate instructors of the Pre-Vet school at my state university. They felt it was perfectly acceptable to experiment with and then terminate the lives of all the animals they utilized, which I found revoltingly unacceptable to my own moral code. After numerous confrontations with these

heartless vivisectionists, I painfully decided to pursue a different career.[101]

In 1966, when moves were being made to pass legislation to protect laboratory animals, the American Veterinary Medical Association testified to congressional committees that while it favored legislation to stop the stealing of pets for subsequent sale to laboratories, it was opposed to the licensing and regulation of research facilities, since this could interfere with research. The basic attitude of the profession was, as an article in the *Journal of the American Veterinary Medical Association* put it, that "the raison d'être of the veterinary profession is the over-all well-being of man—not lower animals."[102] Once the implications of this fine example of speciesism have been grasped, it should surprise no one to learn that veterinarians were part of the experimental teams that performed many of the experiments listed in this chapter. For just one example, look back to the description on p. 27 of the Primate Equilibrium Platform experiment involving exposure to the nerve gas agent, soman. The report from which this description is drawn states: "Routine care of the animals was provided by the Veterinary Sciences Division, USAF School of Aerospace Medicine."

Throughout America, veterinarians are standing by providing "routine care" for animals who are being needlessly abused. Is this what the veterinary profession stands for? (There is some hope for the vets, however, because a new organization of veterinarians has been established to provide support for practitioners and students with ethical concerns about the treatment of nonhuman animals.[103])

Once a pattern of animal experimentation becomes the accepted mode of research in a particular field, the process is self-reinforcing and difficult to break out of. Not only publications and promotions but also the awards and grants that finance research become geared to animal experiments. A proposal for a new experiment with animals is something that the administrators of research funds will be ready to support, if they have in the past supported other experiments on animals. New methods that do not make use of animals will seem less familiar and will be less likely to receive support.

All this helps to explain why it is not always easy for people outside the universities to understand the rationale for the research carried out under university auspices. Originally, perhaps, scholars and researchers just set out to solve the most important problems and did not allow themselves to be influenced by other considerations. No doubt some are still motivated by these concerns. Too often, though, academic research gets bogged down in petty and insignificant details because the big questions have been studied already and they have either been solved or proven too difficult. So the researchers turn away from the well-plowed fields in search of new territory where whatever they find will be new, although the connection with a major problem may be remote. It is not uncommon, as we have seen, for experimenters to admit that similar experiments have been done many times before, but without this or that minor variation; and the most common ending to a scientific publication is "further research is necessary."

When we read reports of experiments that cause pain and are apparently not even intended to produce results of real significance, we are at first inclined to think that there must be more to what is being done than we can understand—that the scientists must have some better reason for what they are doing than their reports indicate. When I describe such experiments to people or quote directly from the researchers' own published reports, the most common reaction I get is puzzlement and skepticism. When we go more deeply into the subject, however, we find that what appears trivial on the surface very often really is trivial. Experimenters themselves often unofficially admit this. H. F. Harlow, whose experiments we encountered at the beginning of this chapter, was for twelve years the editor of the *Journal of Comparative and Physiological Psychology*, a journal that has published more reports of painful experiments on animals than almost any other. At the end of this period, in which Harlow estimated he reviewed about 2,500 manuscripts submitted for publication, he wrote, in a semihumorous farewell note, that "most experiments are not worth doing and the data attained are not worth publishing."[104]

We shouldn't be surprised by this. Researchers, even those in psychology, medicine, and the biological sciences, are human beings and are susceptible to the same influences as any other human beings. They like to get on in their careers, to be promoted, and to have their work read and discussed by their colleagues. Publishing papers in the appropriate journals is an important element in the rise up the ladder of promotion and increased prestige. This happens in every field, in philosophy or history as much as in psychology or medicine, and it is entirely understandable and in itself hardly worth criticizing. The philosophers and historians who publish to improve their career prospects do little harm beyond wasting paper and boring their colleagues; those whose work involves experimenting on animals, however, can cause severe pain or prolonged suffering. Their work should therefore be subject to much stricter standards of necessity.

The government agencies in the United States, Britain, and elsewhere that promote research in the biological sciences have become the major backers of experiments on animals. Indeed, public funds, derived from taxation, have paid for the vast majority of the experiments described in this chapter. Many of these agencies are paying for experiments that have only the remotest connections with the purposes for which the agencies were set up. In the preceding pages I have described experiments that were funded by the United States National Institutes of Health, the Alcohol, Drug Abuse and Mental Health Administration, the Federal Aviation Administration, the Defense Department, the National Science Foundation, the National Aeronautics and Space Administration, and others. It is not easy to understand why the U.S. Army should be paying for a study of the urine spreading patterns of heated, drugged rats or why the U.S. Public Health Service should wish to give out money so that elephants can be given LSD.

Since these experiments are paid for by government agencies, it is hardly necessary to add that there is no law that prevents the scientist from carrying them out. There are laws that prevent ordinary people from beating their dogs to death, but in the United States scientists can do the same thing with impunity, and with no one to check whether their doing so is likely to lead to benefits that would not occur from an ordinary beating. The reason for this is that the strength and prestige of the scientific establishment, supported by the various interest groups—including those who breed animals for sale to laboratories—have been sufficient to stop attempts at effective legal control.

Robert J. White of the Cleveland Metropolitan General Hospital is an experimenter who has specialized in transplanting the heads of monkeys and keeping these monkey heads alive in fluid after they have been totally detached from their bodies. He is a perfect example of the scientist who thinks of a laboratory animal as a "tool for research"—in fact he has himself said that the main purpose of his work on decapitated monkey heads is "to offer a living laboratory tool" for research on the brain. The reporter to whom he made this statement found the visit to White's laboratory "a rare and chilling glimpse into the cold, clinical world of the scientist, where the life of an animal has no meaning beyond the immediate purpose of experimentation."[10]

In White's view, "the inclusion of animals in our ethical system is philosophically meaningless and operationally impossible."[106] In other words, White sees himself as under no ethical constraints in regard to what he does to animals. Hence it is not surprising that another reporter interviewing him should have found that White "chafes at regulations, whether from hospital administrators or insurers. 'I'm an elitist,' he says. He believes doctors should be governed by their peers."[107]

Another active opponent of government regulations is David Baltimore, a professor at Massachusetts Institute of Technology and a Nobel laureate. In a recent address to the national meeting of the American Association for the Advancement of Science he referred to the "long hours" that he and his colleagues had spent fighting regulation of their research.[108] The basis for Baltimore's opposition to such regulation was made clear some years earlier, when he appeared on a television program with Harvard philosopher Robert Nozick and other scientists. Nozick asked the scientists whether the fact that an experiment will kill hundreds of animals is ever regarded, by scientists, as a reason for not performing it. One of the scientists answered: "Not that I know of." Nozick pressed his question: "Don't the animals count at all?" A scientist countered: "Why should they?" At this point Baltimore interjected that he did not think that experimenting on animals raised a moral issue at all.[109]

Men like White and Baltimore may be brilliant scientists, but their utterances on animals show that they are philosophical ignoramuses. I know of not a single professional philosopher writing today who would agree that it is "meaningless" or "impossible" to include animals in our ethical system or that experimenting on animals raises no moral issue. Such statements are, in philosophy, comparable to maintaining that the earth is flat.

American scientists have, so far, been extraordinarily intransigent about public oversight of what they do to animals. They have been successful in squelching even minimal regulations to protect animals from suffering in experiments. In the United States, the only federal law on the matter is the Animal Welfare Act. The law sets the standards for the transportation, housing, and handling of animals sold as pets, exhibited, or intended for use in research. So far as actual experimentation is concerned, however, it allows the researchers to do exactly as they please. This is quite deliberate: the reason given by the U.S. Congress Conference Committee when the act was passed was

> to provide protection for the researcher in this matter by exempting from regulations all animals during actual research or experimentation.... It is not the intention of the committee to interfere in any way with research or experimentation.[110]

One section of the law requires that those private businesses and other organizations that register under the act (neither government agencies doing research nor many smaller facilities have to register) must file a report stating that when painful experiments were performed without the use of pain-relieving drugs, this was necessary to achieve the objectives of the research project. No attempt is made to assess whether these "objectives" are sufficiently important to justify inflicting pain. Under these circumstances the requirement does no more than make additional paperwork, and this is a major complaint among experimenters. They can't, of course, give dogs the continual electric shocks that will produce a state of helplessness if they anesthetize them at the same time; nor can they produce depression in monkeys while keeping them happy or oblivious with drugs. So in such cases they can truthfully state that the objectives of the experiment cannot be achieved if pain-relieving drugs are used, and then go on with the experiment as they would have done before the act came into existence.

So we should not be surprised that, for instance, the report of the Primate Equilibrium Platform experiment with soman should be prefaced with the following statement:

> The animals involved in this study were procured, maintained, and used in accordance with the Animal Welfare Act and the "Guide for the Care and Use of Laboratory Animals" prepared by the Institute of Laboratory Animal Resources–National Research Council.

In fact the same statement appears on the Brooks Air Force Base Training manual for the Primate Equilibrium Platform, on the report of the Armed Forces Radiobiology Research Institute's "primate activity wheel" experiment, and on many other recent American publications from which I have quoted. The statement tells us nothing at all about how much the animals suffered, nor about how trivial the purpose for which they suffered may have been; but it tells us a great deal about the value of the Animal Welfare Act and of the "Guide for the Care and Use of Laboratory Animals" prepared by the National Research Council's Institute of Laboratory Animal Resources.

The complete absence of effective regulation in the United States is in sharp contrast to the situation in many other developed nations. In Britain, for example, no experiment can be conducted without a license granted by the secretary of state for home affairs, and the Animals (Scientific Procedures) Act, 1986, expressly directs that in determining whether to grant a license for an experimental project, "the Secretary of State shall weigh the likely adverse effects on the animals concerned against the benefit likely to accrue." In Australia, the Code of Practice developed by the leading governmental scientific bodies (equivalent to the National Institutes of Health in the United States) requires that all experiments must be approved by an Animal Experimentation Ethics Committee. These committees must include a person with an interest in animal welfare who is not employed by the institution conducting the experiment, and an additional independent person not involved in animal experimentation. The committee must apply a detailed set of principles and conditions that include an instruction to weigh the scientific or educational value of the experiment against the potential effects on the welfare of animals. In addition, anesthesia must be used if the experiment "may cause pain of a kind and degree for which anesthesia would normally be used in medical or veterinary practice." The Australian Code of Practice applies to all researchers obtaining government grants, and under state law is binding on all experimenters in Victoria, New South Wales, and South Australia.[111] Sweden also requires experiments to be approved by committees that include lay members. In 1986, after surveying the laws in Australia, Canada, Japan, Denmark, Germany, the Netherlands, Norway, Sweden, Switzerland, and the United Kingdom, the U.S. Congress Office of Technology Assessment concluded:

> Most of the countries examined for this assessment have laws far more protective of experimental animals than those in the United States. Despite these protections, animal welfare advocates have been applying considerable pressure for even stronger laws, and many countries, including Australia, Switzerland, West Germany, and the United Kingdom, are considering major changes.[112]

Stronger laws have in fact already been passed in Australia and the United Kingdom since that statement was made.

I hope this comparison will not be misunderstood. It is not intended to show that all is well with animal experimentation in countries like the United Kingdom and Australia. That would be far from the truth. In those countries the "balancing" of potential benefits against harm to the animals is still carried out within the assumption of a speciesist attitude to animals, thus rendering it impossible for the interests of animals to be given equal consideration with similar interests of humans. I have compared the situation in the United States with that in other countries only in order to show that American standards in this matter are abysmal, not just by the standards of animal liberationists, but by those accepted by the scientific communities of other major developed nations. It would be salutary for United States scientists to see themselves as their colleagues in other countries see them.

At medical and scientific conferences I attend in Europe and Australia, I am frequently taken aside by scientists who tell me that they may not agree with all my views about animal experimentation, but . . . and then they tell me, with genuine horror in their voice, about something they saw during their last trip to the United States. No wonder that in the respected British science magazine *New Scientist*, a writer recently described the United States as "a country which, as reflected in its legislation to protect animals, seems to be a nation of barbarians."[113] As the United States lagged behind the civilized world in outlawing human slavery, so the United States now lags behind in softening the unrestrained brutalities of animal slavery.

Minor amendments to the United States Animal Welfare Act in 1985 improved exercise requirements for dogs and housing for primates, but failed to deal with the real issue of control over what happens during an experiment. The amendments set up institutional animal committees, but in keeping with the unchanged exemption from interference given to the experiments themselves, these committees have no authority over what goes on in the experiments.[114]

In any case, despite the fact that the Animal Welfare Act was passed more than twenty years ago, its enforcement is virtually nil. For a start, the secretary of agriculture has never even issued regulations extending the act's provisions to mice, rats, birds, and farm animals used in research. Presumably this is because the Department of Agriculture does not even have enough inspectors to check on the conditions of such animals as dogs, cats, and monkeys, let alone birds, rats, mice, and farm animals. As the Office of Technology Assessment said, "funds and personnel for enforcement have never lived up to the expectations of those who believe the primary mission of the existing law to be the prevention or alleviation of experimental animal suffering." OTA staff checked one list of 112 testing facilities, and found that 39 percent were not even registered with the branch of the Department of Agriculture that inspects laboratories. Moreover, the OTA report states that this is probably a conservative estimate of the real number of unregistered, and hence totally uninspected and uncontrolled, animal laboratories.[115]

The United States regulation of animal experimentation is now a continuing farce: there is a law that on its face applies to all warm-blooded laboratory animals, but it can be put into effect only by regulations that, in the words of the Office of Technology Assessment, "probably do not affect a substantial percentage of animals used for experimental purposes." The OTA went on to say that this exclusion of many species from the protection of the act "appears to frustrate the intent of Congress and to be beyond the Secretary of Agriculture's statutory authority."[116] These are strong words for the usually restrained OTA—but three years later, nothing at all has been done to change the situation. Indeed, a 1988 report by a blue ribbon panel of American scientists considered, but rejected, a recommendation that the regulations be extended to cover all warm-blooded animals. No reason was given for this rejection: it stands as another example of the obstructionist attitude of United States scientists to the most elementary improvements in the conditions of the animals they use.[117]

So the farce shows no sign of coming to an end. The trouble is that it is decidedly unfunny. There is no reason to believe that rats and mice are less sensitive to pain and suffering, or less in need of minimum standards for housing and transport, than guinea pigs, hamsters, rabbits, or many other animals.

In descriptions of experiments in this chapter up to now, I have limited myself to summarizing the reports written by the experimenters themselves and published in the scientific journals. That evidence cannot be accused of being exaggerated. But because of the total lack of any adequate inspection or scrutiny of what happens in experiments, the reality is often much worse than the published account. This became clear in 1984 in the case of experiments conducted by Thomas Gennarelli at the University of Pennsylvania. The aim of the experiments was to inflict head injuries on monkeys, and then examine the nature of the damage to the brain. According to the official grant documents the monkeys were to be anesthetized before receiving the head injury. Thus it would seem that the experiments involved no suffering. But members of a group called the Animal Liberation Front had other information. They had also learned that Gennarelli videotaped his experiments. They broke into the laboratory and stole the tapes. When they viewed them, they saw conscious, unanesthetized baboons struggling as they were being strapped down before the head injuries were inflicted. They saw animals writhing, apparently coming out of anesthesia, as surgeons were operating on their exposed brains. They also heard the experimenters mocking and laughing at frightened, suffering animals. The videotapes were so damning that—though it took more than a year of hard work by the Washington-based group People for the Ethical Treatment of Animals and hundreds of animal activists—the secretary of health and human services stopped Gennarelli's funding.[118] Since then, other examples have come to light, based usually on information provided by someone working in the laboratory who has blown the whistle, at the cost of his or her job. In 1986, for instance, Leslie Fain, an animal care technician at Gillette's testing laboratory in Rockville, Maryland, resigned her job and gave Animal Liberationists photos she had taken inside the laboratory. The photos showed Gillette testing new formulations of pink and brown inks for its Paper Mate pens by putting them in the eyes of conscious rabbits. The inks turned out to be extremely irritating, and caused a bloody discharge from the eye in some rabbits.[119] One can only guess at how many laboratories there are in which the abuse of animals is just as bad, but no one has been courageous enough to do anything about it.

When are experiments on animals justifiable? Upon learning of the nature of many of the experiments carried out, some people react by saying that all experiments on animals should be prohibited immediately. But if we make our demands as absolute as this, the experimenters have a ready reply: Would we be prepared to let thousands of humans die if they could be saved by a single experiment on a single animal?

This question is, of course, purely hypothetical. There has never been and never could be a single experiment that saved thousands of lives. The way to reply to this hypothetical question is to pose another: Would the experimenters be prepared to carry out their experiment on a human orphan under six months old if that were the only way to save thousands of lives?

If the experimenters would not be prepared to use a human infant then their readiness to use nonhuman animals reveals an unjustifiable form of discrimination on the basis of species, since adult apes, monkeys, dogs, cats, rats, and other animals are more aware of what is happening to them, more self-directing, and, so far as we can tell, at least as sensitive to pain as a human infant. (I have specified that the human infant be an orphan, to avoid the complications of the feelings of parents. Specifying the case in this way is, if anything, overgenerous to those defending the use of nonhuman animals in experiments, since mammals intended for experimental use are usually separated from their mothers at an early age, when the separation causes distress for both mother and young.)

So far as we know, human infants possess no morally relevant characteristic to a higher degree than adult nonhuman animals, unless we are to count the infants' potential as a characteristic that makes it wrong to experiment on them. Whether this characteristic should count is controversial—if we count it, we shall have to condemn abortion along with experiments on infants, since the potential of the infant and the fetus is the same. To avoid the complexities of this issue, however, we can alter our original question a little and assume that the infant is one with irreversible brain damage so severe as to rule out any mental development beyond the level of a six-month-old infant. There are, unfortunately, many such human beings, locked away in special wards throughout the country, some of them long since abandoned by their parents and other relatives, and, sadly, sometimes unloved by anyone else. Despite their mental deficiencies, the anatomy and physiology of these infants are in nearly all respects identical with those of normal humans. If, therefore, we were to force-feed them with large quantities of floor polish or drip concentrated solutions of cosmetics into their eyes, we would have a much more reliable indication of the safety of these products for humans than we now get by attempting to extrapolate the results of tests on a variety of other species. The LD50 tests, the Draize eye tests, the radiation experiments, the heatstroke experiments, and many others described earlier in this chapter could have told us more about human reactions to the experimental situation if they had been carried out on severely brain-damaged humans instead of dogs or rabbits.

So whenever experimenters claim that their experiments are important enough to justify the use of animals, we should ask them whether they would be prepared to use a brain-damaged human being at a similar mental level to the animals they are planning to use. I cannot imagine that anyone would seriously propose carrying out the experiments described in this chapter on brain-damaged human beings. Occasionally it has become known that medical experiments have been performed on human beings without their consent; one case did concern institutionalized intellectually disabled children, who were given hepatitis.[120] When such harmful experiments on human beings become known, they usually lead to an outcry against the experimenters, and rightly so. They are, very often, a further example of the arrogance of the research worker who justifies everything on the grounds of increasing knowledge. But if the experimenter claims that the experiment is important enough to justify inflicting suffering on animals, why is it not important enough to justify inflicting suffering on humans at the same mental level? What difference is there between the two? Only that one is a member of our species and the other is not? But to appeal to that difference is to reveal a bias no more defensible than racism or any other form of arbitrary discrimination.

The analogy between speciesism and racism applies in practice as well as in theory in the area of experimentation. Blatant speciesism leads to painful experiments on other species, defended on the grounds of their contribution to knowledge and possible usefulness for our species. Blatant racism has led to painful experiments on other races, defended on the grounds of their contribution to knowledge and possible usefulness for the experimenting race. Under the Nazi regime in Germany, nearly two hundred doctors, some of them eminent in the world of medicine, took part in experiments on Jews and Russian and Polish prisoners. Thousands of other physicians knew of these experiments, some of which were the subject of lectures at medical academies. Yet the records show that the doctors sat through verbal reports by doctors on how horrible injuries were inflicted on these "lesser races," and then proceeded to discuss the medical lessons to be learned from them, without anyone making even a mild protest about the nature of the experiments. The parallels between this attitude and that of experimenters today toward animals are striking. Then, as now, subjects were frozen, heated, and put in decompression chambers. Then, as now, these events were written up in dispassionate scientific jargon. The following paragraph is taken from a report by a Nazi scientist of an experiment on a human being, placed in a decompression chamber:

> After five minutes spasms appeared; between the sixth and tenth minute respiration increased in frequency, the TP [test person] losing consciousness. From the eleventh to the thirtieth minute respiration slowed down to three inhalations per minute, only to cease entirely at the end of that period.... About half an hour after breathing ceased, an autopsy was begun.[121]

Decompression chamber experimentation did not stop with the defeat of the Nazis. It shifted to nonhuman animals. At the University of Newcastle on Tyne, in England, for instance, scientists used pigs. The pigs were subjected to up to eighty-one periods of decompression over a period of nine months. All suffered attacks of decompression sickness, and some died from these attacks.[122] The example illustrates only too well what the great Jewish writer Isaac Bashevis Singer has written: "In their behavior towards creatures, all men [are] Nazis."[123]

Experimentation on subjects outside the experimenters' own group is a story that constantly repeats itself with different victims. In the United States the most notorious twentieth-century instance of human experimentation was the deliberate nontreatment of syphilis patients at Tuskegee, Alabama, so that the natural course of the disease could be observed. This was continued long after penicillin was shown to be an effective treatment for syphilis. The untreated victims of the experiment were, of course, blacks.[124] Perhaps the major international human experimentation scandal of the past decade came to light in New Zealand in 1987. A respected doctor at a leading Auckland hospital decided not to treat patients with early signs of cancer. He was trying to prove his unorthodox theory that this form of cancer would not develop, but he did not tell the patients that they were part of an experiment. His theory was wrong, and twenty-seven of his patients died. This time the victims were women.[125]

When such events come to light, the public reaction makes it clear that our sphere of moral concern is wider than that of the Nazis, and we are no longer prepared to countenance a lesser degree of concern for other human beings; but there are still many sentient beings for whom we appear to have no real concern at all.

We have still not answered the question of when an experiment might be justifiable. It will not do to say "Never!" Putting morality in such black-and-white terms is appealing, because it eliminates the need to think about particular cases; but in extreme circumstances, such absolutist answers always break down. Torturing a human being is almost always wrong, but it is not absolutely wrong. If torture were the only way in which we could discover the location of a nuclear bomb hidden in a New York City basement and timed to go off within the hour, then torture would be justifiable. Similarly, if a single experiment could cure a disease like leukemia, that experiment would be justifiable. But in actual life the benefits are always more remote, and more often than not they are nonexistent. So how do we decide when an experiment is justifiable?

We have seen that experimenters reveal a bias in favor of their own species whenever they carry out experiments on nonhumans for purposes that they would not think justified them in using human beings, even brain-damaged ones. This principle gives us a guide toward an answer to our question. Since a speciesist bias, like a racist bias, is unjustifiable, an experiment cannot be justifiable unless the experiment is so important that the use of a brain-damaged human would also be justifiable.

This is not an absolutist principle. I do not believe that it could never be justifiable to experiment on a brain-damaged human. If it really were possible to save several lives by an experiment that would take just one life, and there were no other way those lives could be saved, it would be right to do the experiment. But this would be an extremely rare case. Certainly none of the experiments described in this chapter could pass this test. Admittedly, as with any dividing line, there would be a gray area where it was difficult to decide if an experiment could be justified. But we need not get distracted by such considerations now. As this chapter has shown, we are in the midst of an emergency in which appalling suffering is being inflicted on millions of animals for purposes that on any impartial view are obviously inadequate to justify the suffering. When we have ceased to carry out all those experiments, then there will be time enough to discuss what to do about the remaining ones which are claimed to be essential to save lives or prevent greater suffering.

In the United States, where the present lack of control over experimentation allows the kinds of experiments described in the preceeding pages, a minimal first step would be a requirement that no experiment be conducted without prior approval from an ethics committee that includes animal welfare representatives and is authorized to refuse approval to experiments when it does not consider that the potential benefits outweigh the harm to the animals. As we have seen, systems of this kind already exist in countries such as Australia and Sweden and are accepted as fair and reasonable by the scientific community there. On the basis of the ethical arguments in this book, such a system falls far short of the ideal. The animal welfare representatives on such committees come from groups that hold a spectrum of views, but, for obvious reasons, those who receive and accept invitations to join animal experimentation ethics committees tend to come from the less radical groups within the movement. They may not themselves regard the interests of nonhuman animals as entitled to equal consideration with the interests of humans; or if they do hold such a position, they may find it impossible to put it into practice when judging applications to perform animal experiments, because they would be unable to persuade other members of the committee. Instead, they are likely to insist on proper consideration of alternatives, genuine efforts to minimize pain, and a clear demonstration of significant potential benefits, sufficiently important to outweigh any pain or suffering that cannot be eliminated from the experiment. An animal experimentation ethics committee operating today would almost inevitably apply these standards in a speciesist manner, weighing animal suffering more lightly than potential comparable human benefit; even so, an emphasis on such standards would eliminate many painful experiments now permitted and would reduce the suffering caused by others.

In a society that is fundamentally speciesist, there is no quick solution to such difficulties with ethics committees. For this reason some Animal Liberationists will have nothing to do with them. Instead they demand the total and immediate elimination of all animal experimentation. Such demands have been put forward many times during the last century and a half of anti-vivisection activity, but they have shown no sign of winning over the majority of voters in any country. Meanwhile the number of animals suffering in laboratories continued to grow, until the recent breakthroughs described earlier in this chapter. These breakthroughs resulted from the work of people who found a way around the "all or nothing" mentality that had effectively meant "nothing" as far as the animals were concerned.

One reason the demand for immediate abolition of animal experimentation has failed to persuade the public is that experimenters respond that to accept this demand is to give up the prospect of finding a cure for major diseases that still kill us and our children. In the United States, where experimenters can do virtually as they please with animals, one way of making progress might be to ask those who use this argument to defend the need for animal experiments whether they would be prepared to accept the verdict of an ethics committee that, like those in many other countries, includes animal welfare representatives and is entitled to weigh the costs to the animals against the possible benefits of the research. If the answer is no, the defense of animal experimentation by reference to the need to cure major diseases has been proved to be simply a deceitful distraction that serves to mislead the public about what the experimenters want: permission to do whatever they like with animals. For otherwise why would the experimenter not be prepared to leave the decision on carrying out the experiment to an ethics committee, which would surely be as keen to see major diseases ended as the rest of the community? If the answer is yes, the experimenter should be asked to sign a statement asking for the creation of such an ethics committee.

Suppose that we were able to go beyond minimal reforms of the sort that already exist in the more enlightened nations. Suppose we could reach a point at which the interests of animals really were given equal consideration with the similar interests of human beings. That would mean the end of the vast industry of animal experimentation as we know it today. Around the world, cages would empty and laboratories would close down. It should not be thought, though, that medical research would grind to a halt or that a flood of untested products would come on to the market. So far as new products are concerned it is true, as I have already said, that we would have to make do with fewer of them, using ingredients already known to be safe. That does not seem to be any great loss. But for testing really essential products, as well as for other kinds of research, alternative methods not requiring animals can and would be found.

In the first edition of this book I wrote that "scientists do not look for alternatives simply because they do not care enough about the animals they are using." Then I made a prediction: "Considering how little effort has been put into this field, the early results promise much greater progress if the effort is stepped up." In the past decade, both these statements have proved true. We have already seen that in product testing there has been a huge increase in the amount of effort put into looking for alternatives to animal experiments—not because scientists have suddenly started to care more about animals, but as a result of hard-fought campaigns by Animal Liberationists. The same thing could happen in many other fields of animal experimentation.

Although tens of thousands of animals have been forced to inhale tobacco smoke for months and even years, the proof of the connection between tobacco use and lung cancer was based on data from clinical observations in human beings.[126] The United States government continues to pour billions of dollars into research on cancer, while it also subsidizes the tobacco industry. Much of the research money goes toward animal experiments, many of them only remotely connected with fighting cancer—experimenters have been known to relabel their work "cancer research" when they found they could get more money for it that way than under some other label. Meanwhile we are continuing to lose the fight against most forms of cancer. Figures released in 1988 by the United States National Cancer Institute show that the overall rate of cancer, even when adjusted for the increasing age of the population, has been rising at about 1 percent per year for thirty years. Recent reports of a decline in lung cancer rates among younger Americans may be the first sign of a reversal in this trend, since lung cancer causes more deaths than any other form of cancer. If lung cancer is declining, however, this welcome news is not the result of any improvement in treatment but of younger people, especially white males, smoking less. Lung cancer survival rates have scarcely changed.[127] We know that smoking causes between 80 and 85 percent of all lung cancer cases. We must ask ourselves: Can we justify forcing thousands of animals to inhale cigarette smoke so that they develop lung cancer, when we know we could virtually wipe out the disease by eliminating the use of tobacco? If people decide to continue to smoke, knowing that by doing so they risk lung cancer, is it right to make animals suffer the cost of this decision?

Our poor record in the treatment of lung cancer is matched in cancer treatment more generally. Although there have been successes in treating some specific cancers, since 1974 the number of people surviving for five years or more after cancer has been diagnosed has increased by less than 1 percent.[128] Prevention, particularly through educating people to lead healthier lives, is a more promising approach.

More and more scientists are now appreciating that animal experimentation often actually hinders the advance of our understanding of diseases in humans and their cure. For example, researchers at the National Institute of Environmental Health Sciences, in North Carolina, recently warned that animal tests may fail to pick up chemicals that cause cancer in people. Exposure to arsenic seems to increase the risk that a person will develop cancer, but it does not have this effect in laboratory tests on animals.[19] A malaria vaccine developed in the United States in 1985 at the prestigious Walter Reed Army Institute of Research worked in animals, but proved largely ineffective in humans; a vaccine developed by Colombian scientists working with human volunteers has proven more effective.[130] Nowadays defenders of animal research often talk about the importance of finding a cure for AIDS; but Robert Gallo, the first American to isolate HIV (the AIDS virus), has said that a potential vaccine developed by the French researcher Daniel Zagury had shown itself to be more effective in stimulating HIV antibody production in human beings than in animals; and he added: "The results in chimps haven't been too exciting.... Maybe we should go into testing in man more aggressively."[131] Significantly, people with AIDS have endorsed this call: "Let us be your guinea pigs," pleaded gay activist Larry Kramer.[132] Obviously this plea makes sense. A cure will be found faster if experimentation is done directly on human volunteers; and because of the nature of the disease, and the strong bonds between many members of the gay community, there is no shortage of volunteers. Care needs to be taken, of course, that those volunteering genuinely understand what they are doing and are under no pressure or coercion to take part in an experiment. But it would not be unreasonable to give such consent. Why should people be dying from an invariably fatal disease while a potential cure is tested on animals who do not normally develop AIDS anyway?

The defenders of animal experimentation are fond of telling us that animal experimentation has greatly increased our life expectancy. In the midst of the debate over reform of the British law on animal experimentation, for example, the Association of the British Pharmaceutical Industry ran a full-page advertisement in the *Guardian* under the headline "They say life begins at forty. Not so long ago, that's about when it ended." The advertisement went on to say that it is now considered to be a tragedy if a man dies in his forties, whereas in the nineteenth century it was commonplace to attend the funeral of a man in his forties, for the average life expectancy was only forty-two. The advertisement stated that "it is thanks largely to the breakthroughs that have been made through research which requires animals that most of us are able to live into our seventies."

Such claims are simply false. In fact, this particular advertisement was so blatantly misleading that a specialist in community medicine, Dr. David St. George, wrote to *The Lancet* saying "the advertisement is good teaching material, since it illustrates two major errors in the interpretation of statistics." He also referred to Thomas McKeown's influential book *The Role of Medicine*, published in 1976,[133] which set off a debate about the relative contributions of social and environmental changes, as compared with medical intervention, in improvements in mortality since the mid-nineteenth century; and he added:

> This debate has been resolved, and it is now widely accepted that medical interventions had only a marginal effect on population mortality and mainly at a very late stage, after death rates had already fallen strikingly.[134]

J. B. and S. M. McKinley reached a similar conclusion in a study of the decline of ten major infectious diseases in the United

States. They showed that in every case except poliomyelitis the death rate had already fallen dramatically (presumably because of improved sanitation and diet) before any new form of medical treatment was introduced. Concentrating on the 40 percent fall in crude mortality in the United States between 1910 and 1984, they estimated "conservatively" that

> perhaps 3.5 percent of the fall in the overall death rate can be explained through medical interventions for the major infectious diseases. Indeed, given that it is precisely for these diseases that medicine claims most success in lowering mortality, 3.5 percent probably represents a reasonable upper-limit estimate of the total contribution of medical measures to the decline in infectious disease mortality in the United States.[135]

Remember that this 3.5 percent is a figure for all medical intervention. The contribution of animal experimentation itself can be, at most, only a fraction of this tiny contribution to the decline in mortality.

No doubt there are some fields of scientific research that will be hampered by any genuine consideration of the interests of animals used in experimentation. No doubt there have been some advances in knowledge which would not have been attained as easily without using animals. Examples of important discoveries often mentioned by those defending animal experimentation go back as far as Harvey's work on the circulation of blood. They include Banting and Best's discovery of insulin and its role in diabetes; the recognition of poliomyelitis as a virus and the development of a vaccine for it; several discoveries that served to make open heart surgery and coronary artery bypass graft surgery possible; and the understanding of our immune system and ways to overcome rejection of transplanted organs.[136] The claim that animal experimentation was essential in making these discoveries has been denied by some opponents of experimentation.[137] I do not intend to go into the controversy here. We have just seen that any knowledge gained from animal experimentation has made at best a very small contribution to our increased lifespan; its contribution to improving the quality of life is more difficult to estimate. In a more fundamental sense, the controversy over the benefits derived from animal experimentation is essentially unresolvable, because even if valuable discoveries were made using animals, we cannot say how successful medical research would have been if it had been compelled, from the outset, to develop alternative methods of investigation. Some discoveries would probably have been delayed, or perhaps not made at all; but many false leads would also not have been pursued, and it is possible that medicine would have developed in a very different and more efficacious direction, emphasizing healthy living rather than cure.

In any case, the ethical question of the justifiability of animal experimentation cannot be settled by pointing to its benefits for us, no matter how persuasive the evidence in favor of such benefits may be. The ethical principle of equal consideration of interests will rule out some means of obtaining knowledge. There is nothing sacred about the right to pursue knowledge. We already accept many restrictions on scientific enterprise. We do not believe that scientists have a general right to perform painful or lethal experiments on human beings without their consent, although there are many cases in which such experiments would advance knowledge far more rapidly than any other method. Now we need to broaden the scope of this existing restriction on scientific research.

Finally, it is important to realize that the major health problems of the world largely continue to exist, not because we do not know how to prevent disease and keep people healthy, but because no one is putting enough effort and money into doing what we already know how to do. The diseases that ravage Asia, Africa, Latin America, and the pockets of poverty in the industrialized West are diseases that, by and large, we know how to cure. They have been eliminated in communities that have adequate nutrition, sanitation, and health care. It has been estimated that 250,000 children die each week around the world, and that one quarter of these deaths are by dehydration caused by diarrhea. A simple treatment, already known and needing no animal experimentation, could prevent the deaths of these children.[138] Those who are genuinely concerned about improving health care would probably make a more effective contribution to human health if they left the laboratories and saw to it that our existing stock of medical knowledge reached those who need it most.

When all this has been said, there still remains the practical question: What can be done to change the widespread practice of experimenting on animals? Undoubtedly, some action that will change government policies is needed, but what action precisely? What can the ordinary citizen do to help bring about change?

Legislators tend to ignore protests about animal experimentation from their constituents, because they are overly influenced by scientific, medical, and veterinary groups. In the United States, these groups maintain registered political lobbies in Washington, and they lobby hard against proposals to restrict experimentation. Since legislators do not have the time to acquire expertise in these fields, they rely on what the "experts" tell them. But this is a moral question, not a scientific one, and the "experts" usually have an interest in the continuation of experimentation or else are so imbued with the ethic of furthering knowledge that they cannot detach themselves from this stance and make a critical examination of what their colleagues do. Moreover, professional public relations organizations have now emerged, such as the National Association for Biomedical Research, whose sole purpose is to improve the image of animal research with the public and with legislators. The association has published books, produced videotapes, and conducted workshops on how researchers should defend experimentation. Along with a number of similar organizations, it has prospered as more people have become concerned about the experimentation issue. We have already seen, in the case of another lobby group, the Association of the British Pharmaceutical Industry, how such groups can mislead the public. Legislators must learn that when discussing animal experimentation they have to treat these organizations, and also the medical, veterinary, psychological, and biological associations, as they would treat General Motors and Ford when discussing air pollution.

Nor is the task of reform made any easier by the large companies involved in the profitable businesses of breeding or trapping animals and selling them, or manufacturing and marketing the cages for them to live in, the food used to feed them, and the equipment used to experiment on them. These companies are prepared to spend huge amounts of money to oppose legislation that will deprive them of their profitable markets. With financial interests like these allied to the prestige of medicine and science the struggle to end speciesism in the laboratory is bound to be difficult and protracted. What is the best way to make progress?

It does not seem likely that any major Western democracy is going to abolish all animal experimentation at a stroke. Governments just do not work like that. Animal experimentation will only be ended when a series of piecemeal reforms have reduced its importance, led to its replacement in many fields, and largely changed the public attitude to animals. The immediate task, then, is to work for these partial goals, which can be seen as milestones on the long march to the elimination of all exploitation of sentient animals. All concerned to end animal suffering can try to make known what is happening at universities and commercial laboratories in their own communities. Consumers can refuse to purchase products that have been tested on animals—especially in cosmetics, alternatives are now available. Students should decline to carry out experiments they consider unethical. Anyone can study the academic journals to find out where painful experiments are being carried out, and then find some way of making the public aware of what is happening.

It is also necessary to make the issue political. As we have already seen, legislators receive huge numbers of letters about animal experiments. But it has taken many years of hard work to make animal experimentation a political issue. Fortunately this has now started to happen in several countries. In Europe and Australia animal experimentation is being addressed seriously by the political parties, especially those closer to the Green end of the political spectrum. In the 1988 United States presidential election, the Republican party platform said that the process of certifying alternatives to animal testing of drugs and cosmetics should be made simpler and quicker.

The exploitation of laboratory animals is part of the larger problem of speciesism and it is unlikely to be eliminated altogether until speciesism itself is eliminated. Surely one day, though, our children's children, reading about what was done in laboratories in the twentieth century, will feel the same sense of horror and incredulity at what otherwise civilized people could do that we now feel when we read about the atrocities of the Roman gladiatorial arenas or the eighteenth-century slave trade.

Chapter 3

Down on the Factory Farm . . .

or what happened to your dinner when it was still an animal

For most human beings, especially those in modern urban and suburban communities, the most direct form of contact with non-human animals is at mealtime: we eat them. This simple fact is the key to our attitudes to other animals, and also the key to what each one of us can do about changing these attitudes. The use and abuse of animals raised for food far exceeds, in sheer numbers of animals affected, any other kind of mistreatment. Over 100 million cows, pigs, and sheep are raised and slaughtered in the United States alone each year; and for poultry the figure is a staggering 5 billion. (That means that about eight thousand birds—mostly chickens—will have been slaughtered in the time it takes you to read this page.) It is here, on our dinner table and in our neighborhood supermarket or butcher's shop, that we are brought into direct touch with the most extensive exploitation of other species that has ever existed.

In general, we are ignorant of the abuse of living creatures that lies behind the food we eat. Buying food in a store or restaurant is the culmination of a long process, of which all but the end product is delicately screened from our eyes. We buy our meat and poultry in neat plastic packages. It hardly bleeds. There is no reason to associate this package with a living, breathing, walking, suffering animal. The very words we use conceal its origins: we eat beef, not bull, steer, or cow, and pork, not pig—although for some reason we seem to find it easier to face the true nature of a leg of lamb. The term "meat" is itself deceptive. It originally meant any solid food, not necessarily the flesh of animals. This usage still lingers in an expression like "nut meat," which seems to imply a substitute for "flesh meat" but actually has an equally good claim to be called "meat" in its own right. By using the more general "meat" we avoid facing the fact that what we are eating is really flesh.

These verbal disguises are merely the top layer of a much deeper ignorance of the origin of our food. Consider the images conjured up by the word "farm": a house; a barn; a flock of hens, overseen by a strutting rooster, scratching around the farmyard; a herd of cows being brought in from the fields for milking; and perhaps a sow rooting around in the orchard with a litter of squealing piglets running excitedly behind her.

Very few farms were ever as idyllic as that traditional image would have us believe. Yet we still think of a farm as a pleasant place, far removed from our own industrial, profit-conscious city life. Of those few who think about the lives of animals on farms, not many know much about modern methods of animal raising. Some people wonder whether animals are slaughtered painlessly, and anyone who has followed a truckload of cattle on the road will probably know that farm animals are transported in extremely crowded conditions; but not many suspect that transportation and slaughter are anything more than the brief and inevitable conclusion of a life of ease and contentment, a life that contains the natural pleasures of animal existence without the hardships that wild animals must endure in their struggle for survival.

These comfortable assumptions bear little relation to the realities of modern farming. For a start, farming is no longer controlled by simple country folk. During the last fifty years, large corporations and assembly-line methods of production have turned agriculture into agribusiness. The process began when big companies gained control of poultry production, once the preserve of the farmer's wife. Today, fifty large corporations virtually control all poultry production in the United States. In the field of egg production, where fifty years ago a big producer might have had three thousand laying hens, today many producers have more than 500,000 layers, and the largest have over 10 million. The remaining small producers have had to adopt the methods of the giants or else go out of business. Companies that had no connection with agriculture have become farmers on a huge scale in order to gain tax concessions or to diversify profits. Greyhound Corporation now produces turkeys, and your roast beef may have come from John Hancock Mutual Life Insurance or from one of a dozen oil companies that have invested in cattle feeding, building feedlots that hold 100,000 or more cattle.[1]

The big corporations and those who must compete with them are not concerned with a sense of harmony among plants, animals, and nature. Farming is competitive and the methods adopted are those that cut costs and increase production. So farming is now "factory farming." Animals are treated like machines that convert low-priced fodder into high-priced flesh, and any innovation will be used if it results in a cheaper "conversion ratio." Most of this chapter is simply a description of these methods, and of what they mean for the animals to whom they are applied. The aim is to demonstrate that under these methods animals lead miserable lives from birth to slaughter. Once again, however, my point is not that the people who do these things to the animals are cruel and wicked. On the contrary, the attitudes of the consumers and the producers are not fundamentally different. The farming methods I am about to describe are merely the logical application of the attitudes and prejudices that are discussed elsewhere in this book. Once we place nonhuman animals outside our sphere of moral consideration and treat them as things we use to satisfy our own desires, the outcome is predictable.

As in the previous chapter, in order to make my account as objective as possible I have not based the descriptions that follow on my own personal observations of farms and the conditions on them. Had I done so I could have been charged with writing a selective, biased account, based on a few visits to unusually bad farms. Instead, the account is drawn largely from the sources that can be expected to be most favorable to the farming industry: the magazines and trade journals of the farm industry itself.

Naturally, articles directly exposing the suffering of farm animals are not to be found in farm magazines, especially not now that the sensitivity of the issue has been brought to the industry's attention. Farm magazines are not interested in the question of animal suffering in itself. Farmers are sometimes advised to avoid practices that would make their animals suffer because the animals will gain less weight under these conditions; and they are urged to handle their animals less roughly when they send them to slaughter because a bruised carcass fetches a lower price; but the idea that we should avoid confining animals in uncomfortable conditions simply because this is in itself a bad thing is not mentioned. Ruth Harrison, the author of *Animal Machines*, a pioneering exposé of intensive farming methods in Britain, concluded that "cruelty is acknowledged only where profitability ceases."[2] That, certainly, is the attitude exhibited in the pages of the farming magazines, in the United States as well as in Britain.

Still, we can learn a great deal about the conditions of farm animals from the farm magazines. We learn about the attitudes of some of the farmers to the animals under their absolute and unrestricted rule, and we learn also about the new methods and techniques that are being adopted and about the problems that arise with these techniques. Provided we know a little about the requirements of farm animals, this information is enough to give us a broad picture of animal farming today. We can put the picture in sharper focus by turning to some of the scientific studies of the welfare of farm animals that, in response to the pressure of the Animal Liberation movement, are appearing in increasing numbers in agricultural and veterinary journals.

The first animal to be removed from the relatively natural conditions of the traditional farm was the chicken. Human beings use chickens in two ways: for their flesh and for their eggs. There are now standard mass-production techniques for obtaining both of these products.

Promoters of agribusiness consider the rise of the chicken industry to be one of the great success stories of farming. At the end of World War II chicken for the table was still relatively rare. It came mainly from small independent farmers or from the unwanted males produced by egg-laying flocks. Today in the United States, 102 million broilers—as table chickens are called—are slaughtered each week after being reared in highly automated factorylike plants that belong to the large corporations that control production. Eight of these corporations account for over 50 percent of the 5.3 billion birds killed annually in the U.S.[3]

The essential step in turning chickens from farmyard birds into manufactured items was confining them indoors. A producer of broilers gets a load of 10,000, 50,000, or more day-old chicks from the hatcheries, and puts them into a long, windowless shed—

usually on the floor, although some producers use tiers of cages in order to get more birds into the same size shed. Inside the shed, every aspect of the birds' environment is controlled to make them grow faster on less feed. Food and water are fed automatically from hoppers suspended from the roof. The lighting is adjusted according to advice from agricultural researchers: for instance, there may be bright light twenty-four hours a day for the first week or two, to encourage the chicks to gain weight quickly; then the lights may be dimmed slightly and made to go off and on every two hours, in the belief that the chickens are readier to eat after a period of sleep; finally there comes a point, around six weeks of age, when the birds have grown so much that they are becoming crowded, and the lights will then be made very dim at all times. The point of this dim lighting is to reduce the aggression caused by crowding.

Broiler chickens are killed when they are seven weeks old (the natural lifespan of a chicken is about seven years). At the end of this brief period, the birds weigh between four and five pounds; yet they still may have as little as half a square foot of space per chicken—or less than the area of a sheet of standard typing paper. (In metric terms, this is 450 square centimeters for a hen weighing more than two kilos.) Under these conditions, when there is normal lighting, the stress of crowding and the absence of natural outlets for the birds' energies lead to outbreaks of fighting, with birds pecking at each other's feathers and sometimes killing and eating one another. Very dim lighting has been found to reduce such behavior and so the birds are likely to live out their last weeks in near-darkness.

Feather-pecking and cannibalism are, in the broiler producer's language, "vices." They are not natural vices, however; they are the result of the stress and crowding to which modern broiler producers subject their birds. Chickens are highly social animals, and in the farmyard they develop a hierarchy, sometimes called a "pecking order." Every bird yields, at the food trough or elsewhere, to those who are higher in the pecking order, and takes precedence over those who are below. There may be a few confrontations before the order is established, but more often than not a show of force, rather that actual physical contact, is enough. As Konrad Lorenz, a renowned observer of animal behavior, wrote in the days when flocks were still small:

> Do animals thus know each other among themselves? They certainly do.... Every poultry farmer knows that... there exists a very definite order, in which each bird is afraid of those that are above her in rank. After some few disputes, which need not necessarily come to blows, each bird knows which of the others she has to fear and which must show respect to her. Not only physical strength, but also personal courage, energy, and even the self-assurance of every individual bird are decisive in the maintenance of the pecking order.[4]

Other studies have shown that a flock of up to ninety chickens can maintain a stable social order, each bird knowing its place; but 80,000 birds crowded together in a single shed is obviously a different matter. The birds cannot establish a social order, and as a result they fight frequently with each other. Quite apart from the inability of the individual bird to recognize so many other birds, the mere fact of extreme crowding probably contributes to irritability and excitability in chickens, as it does in human beings and other animals. This is something that farmers have long known:

> Feather-pecking and cannibalism easily become serious vices among birds kept under intensive conditions. They mean lower productivity and lost profits. Birds become bored and peck at some outstanding part of another bird's plumage.... While idleness and boredom are predisposing causes of the vices, cramped, stuffy and overheated housing are contributory causes.[5]

Farmers must stop "vices" since they cost money; but, although they may know that overcrowding is the root cause, they cannot do anything about this, since in the competitive state of the industry, eliminating overcrowding could mean eliminating one's profit margin at the same time. Costs for the building, for the automatic feeding equipment, for the fuel used to heat and ventilate the building, and for the labor would remain the same, but with fewer birds per shed to sell, income would be reduced. So farmers direct their efforts to reducing the consequences of the stress that costs them money. The unnatural way in which the birds are kept causes the vices, but to control them the poultry farmer must make the conditions still more unnatural. Very dim lighting is one way of doing this. A more drastic step, though one now very widely used in the industry, is "debeaking."

First started in San Diego in the 1940s, debeaking used to be performed with a blowtorch. The farmer would burn away the upper beaks of the chickens so that they were unable to pick at each other's feathers. A modified soldering iron soon replaced this crude technique, and today specially designed guillotinelike devices with hot blades are the preferred instrument. The infant chick's beak is inserted into the instrument, and the hot blade cuts off the end of it. The procedure is carried out very quickly, about fifteen birds a minute. Such haste means that the temperature and sharpness of the blade can vary, resulting in sloppy cutting and serious injury to the bird:

> An excessively hot blade causes blisters in the mouth. A cold or dull blade may cause the development of a fleshy, bulb-like growth on the end of the mandible. Such growths are very sensitive.[6]

Joseph Mauldin, a University of Georgia extension poultry scientist, reported on his field observations at a conference on poultry health:

> There are many cases of burned nostrils and severe mutilations due to incorrect procedures which unquestionably influence acute and chronic pain, feeding behavior and production factors. I have evaluated beak trimming quality for private broiler companies and most are content to achieve 70% falling into properly trimmed categories.... Replacement pullets have their beaks trimmed by crews who are paid for quantity rather than quality work.[7]

Even when the operation is done correctly, it is a mistake to think of it as a painless procedure, like cutting toenails. As an expert British government committee under zoologist Professor F. W. Rogers Brambell found some years ago:

> Between the horn and the bone is a thin layer of highly sensitive soft tissue, resembling the "quick" of the human nail.

> The hot knife used in debeaking cuts through this complex of horn, bone and sensitive tissue, causing severe pain.[8]

Moreover the damage done to the bird by debeaking is long term: chickens mutilated in this way eat less and lose weight for several weeks.[9] The most likely explanation for this is that the injured beak continues to cause pain. J. Breward and M. J. Gentle, researchers at the British Agricultural and Food Research Council's Poultry Research Centre, investigated the beak stumps of debeaked hens and found that the damaged nerves grew again, turning in on themselves to form a mass of intertwining nerve fibers, called a neuroma. These neuromas have been shown in humans with amputated stumps to cause both acute and chronic pain. Breward and Gentle found that this is probably also the case in the neuromas formed by debeaking.[10] Subsequently Gentle, expressing himself with the caution to be expected from a poultry scientist writing in a scientific journal, has said:

> In conclusion, it is fair to say that we do not know how much discomfort or pain birds experience after beak trimming but in a caring society they should be given the benefit of the doubt. To prevent cannibalism and feather pecking of poultry, good husbandry is essential and in circumstances where light intensity cannot be controlled the only alternative is to attempt to breed birds which do not exhibit these damaging traits.[11]

There is also another possible solution. Debeaking, which is routinely performed in anticipation of cannibalism by most producers, greatly reduces the amount of damage a chicken can do to other chickens. But it obviously does nothing to reduce the stress and overcrowding that lead to such unnatural cannibalism in the first place. Old-fashioned farmers, keeping a small flock with plenty of space, had no need to debeak their birds.

Once, chickens were individuals; if a chicken bullied others (and this could happen, though it was not the general rule) that bird would be removed from the flock. Similarly, birds who fell sick or were injured could be attended to, or if necessary, quickly killed. Now one person looks after tens of thousands of birds. A United States secretary of agriculture wrote enthusiastically about how one person could care for 60,000 to 75,000 broilers.[12] *Poultry World* recently published a feature story on the broiler unit of David Dereham, who takes care of 88,000 broilers, housed under one roof, all by himself, and farms sixty acres of land as well! "Take care of" does not mean what it used to, since if a poultry farmer were to spend no more than one second a day inspecting each bird, it would take more than twenty-four hours a day merely to complete the inspection of 88,000 birds, let alone do the other chores and a bit of farming on the side. And then there is the exceedingly dim lighting to make the task of inspection even more difficult. In fact, all the modern poultry farmer does is remove dead birds. It is cheaper to lose a few extra birds in this way than to pay for the additional labor needed to watch the health of individual birds.

In order to allow total control of light and some control of temperature (there is usually heating, but rarely cooling) the broiler sheds have solid, windowless walls and rely on artificial ventilation. The birds never see daylight, until the day they are taken out to be killed; nor do they breathe air which is not heavy with the ammonia from their own droppings. The ventilation is adequate to keep the birds alive in normal circumstances, but if there should be a mechanical failure they soon suffocate. Even as obvious a possibility as a power failure can be disastrous, since not all broiler producers have their own auxiliary power units.

Among other ways in which birds can suffocate in a broiler house is a phenomemon known as "piling." Chickens kept in the broiler sheds become nervous, jittery creatures. Unused to strong light, loud noise, or other intrusions, they may panic at a sudden disturbance and flee to one corner of the shed. In their terrified rush to safety they pile on top of each other so that, as one poultry farmer describes it, they "smother each other in a pitiful heap of bodies in one corner of the rearing area."[14]

Even if the birds escape these hazards, they may succumb to any of a number of diseases that are often prevalent in the broiler houses. One new and still mysterious cause of death is known simply as "acute death syndrome," or ADS. Apparently the product of the unnatural conditions generated by the broiler industry, ADS has been shown to kill an average of roughly 2 percent of broiler flocks in Canada and Australia, and presumably the figures are similar wherever the same methods are used.[15] It has been described in the following way:

> Chickens exhibited a sudden attack prior to death characterised by loss of balance, violent flapping, and strong muscular contractions.... Birds were observed to fall forwards or backwards during the initial loss of balance and could turn over on their back or their sternum during the course of violent flapping.[16]

None of the studies offers a clear explanation of why these apparently healthy chickens should suddenly collapse and die, but a poultry specialist with the British Ministry of Agriculture has linked it to the very goal for which the entire broiler industry strives—rapid growth:

> Broiler mortality levels have increased and it is reasonable to speculate whether this can be indirectly attributed to the very considerable genetic and nutritional advances that have been made. In other words, we may be expecting broilers to grow too quickly—multiplying their weight 50–60 times in 7 weeks.... "Flip-overs," that is, the sudden death of thriving young broilers (usually males) may also be connected with this "super-charged" growth.[17]

The fast growth rate also causes crippling and deformities that force producers to kill an additional 1 to 2 percent of broiler chickens—and since only severe cases are culled, the number of birds suffering from deformities is bound to be much higher.[18] The authors of a study of one particular form of crippling concluded: "We consider that birds might have been bred to grow so fast that they are on the verge of structural collapse."[19]

The atmosphere in which the birds must live is itself a health hazard. During the seven or eight weeks the birds are in the sheds, no effort is made to change the litter or remove the birds' droppings. Despite mechanical ventilation, the air becomes charged with ammonia, dust, and microorganisms. Studies have shown that, as one might expect, dust, ammonia, and bacteria have damaging effects on the birds' lungs.[20] The department of community medicine at the University of Melbourne, Australia, conducted a study into the health hazards of this atmosphere for chicken farmers. They found that 70 percent of farmers reported sore eyes, nearly 30 percent regular coughing, and nearly 15 percent asthma and chronic bronchitis. As a result, the researchers warned chicken farmers to spend as little time as possible in their sheds and to wear a respirator when they go in. But the study said nothing about respirators for the chickens.[21]

When the birds must stand and sit on rotting, dirty, ammonia-charged litter, they also suffer from ulcerated feet, breast blisters, and hock burns. "Chicken parts" are often the remaining parts of damaged birds whose bodies cannot be sold whole. Damage to the feet, however, is not a problem for the industry, since the feet are cut off after slaughter anyway.

If living in long, crowded, ammonia-filled, dusty, windowless sheds is stressful, the birds' first and only experience of sunlight is no less so. The doors will be flung open and the birds, accustomed now to semidarkness, are grabbed by the legs, carried out upside down, and summarily stuffed into crates which are piled on the back of a truck. Then they are driven to the "processing" plant where the chickens are to be killed, cleaned, and turned into neat plastic packages. At the plant they are taken off the truck and stacked, still in crates, to await their turn. That may take several hours, during which time they remain without food and water. Finally they are taken out of the crates and hung upside down on the conveyor belt taking them to the knife that will end their joyless existence.

The plucked and dressed bodies of the chickens will then be sold to millions of families who will gnaw on their bones without pausing for an instant to think that they are eating the dead body of a once living creature, or to ask what was done to that creature in order to enable them to buy and eat its body. And if they did stop to ask, where would they find the answer? If they get their information from the chicken tycoon Frank Perdue, the fourth largest broiler producer in the United States, but definitely first in self-promotion, they will be told that the chickens on his "farm" are pampered and "lead such a soft life."[22] How are ordinary people to find out that Perdue keeps his chickens in 150-yard-long buildings that house 27,000 birds? How are they to know that Perdue's mass production system alone kills 6.8 million birds a week, and that, like many other broiler producers, he cuts the beaks off his chickens in order to prevent them from becoming cannibals under the stress of modern factory life?[23]

Perdue's publicity promotes a common myth: that economic rewards for the farmer and a good life for the birds or animals go hand in hand. Apologists for factory farming often say that if the birds or animals were not happy, they would not thrive and hence would not be profitable. The broiler industry provides a clear refutation of this naive myth. A study published in *Poultry Science* showed that giving chickens as little as 372 square centimeters per bird (20 percent less than the standard amount used in the industry) could be profitable, even though so small a space allowance meant that 6.4 percent of the birds died (more than at lower densities), that birds were underweight, and that there was a high incidence of breast blisters. As the authors point out, the key to profitability in the poultry industry is not profit per bird, but profit for the unit as a whole:

> Mean monetary returns per bird started to decline... as stocking density increased. However, when monetary returns were calculated on the basis of returns per unit of floor area, the reverse effect occurred; monetary returns increased as stocking density increased. Although extremely high stocking densities were tested, the point of diminishing returns was not reached despite the reduction in growth rate.[24]

The reader who, after reading this section, is contemplating buying turkey instead of chicken should be warned that this traditional centerpiece of the family's Thanksgiving dinner is now reared by the same methods as broiler chickens and that debeaking is the general rule among turkeys too. According to *Turkey World*, an "explosion of turkey production" has been taking place during the last few years and is expected to continue. The $2 billion turkey industry raised 207 million turkeys in 1985, with twenty large corporations producing over 80 percent of them. Turkeys spend between thirteen and twenty-four weeks in intensive conditions, more than twice as long as their smaller counterparts, before they meet their end.[25]

"A hen," Samuel Butler once wrote, "is only an egg's way of making another egg." Butler, no doubt, thought he was being funny; but when Fred C. Haley, president of a Georgia poultry firm that controls the lives of 225,000 laying hens, describes the hen as "an egg producing machine" his words have more serious implications. To emphasize his businesslike attitude, Haley adds, "The object of producing eggs is to make money. When we forget this objective, we have forgotten what it is all about."[26]

Nor is this only an American attitude. A British farming magazine has told its readers:

> The modern layer is, after all, only a very efficient converting machine, changing the raw material—feedingstuffs—into the finished product—the egg—less, of course, maintenance requirements.[27]

The idea that the layer is an efficient way to turn feed into eggs is common in the industry trade journals, particularly in advertisements. As may be anticipated, its consequences for the laying hens are not good.

Laying hens go through many of the same procedures as broilers, but there are some differences. Like broilers, layers have to be debeaked, to prevent the cannibalism that would otherwise occur in their crowded conditions; but because they live much longer than broilers, they often go through this operation twice. So we find poultry specialist Dick Wells, head of Britain's National Institute of Poultry Husbandry, recommending debeaking "sometime between 5 and 10 days of age," because there is less stress on the chicks at this time than if the operation is done earlier, and in addition "it is a good way of decreasing the risk of early mortality."[28] When the hens are moved from the growing house to the laying facility between twelve and eighteen weeks of age they are often debeaked again.[29]

The sufferings of laying chickens begin early in life. The newly hatched chicks are sorted into males and females by a "chick-puller." Since the male chicks have no commercial value, they are discarded. Some companies gas the little birds, but often they are dumped alive into a plastic sack and allowed to suffocate under the weight of other chicks dumped on top of them. Others are ground up, while still alive, to be turned into feed for their sisters. At least 160 million birds are gassed, suffocated, or die this way every year in the United States alone.[30] Just how many suffer each particular fate is impossible to tell, because no records are kept: the growers think of getting rid of male chicks as we think of putting out the trash.

Life for the female laying birds is longer, but this is scarcely a benefit. Pullets (as the younger birds not yet ready to lay are called) used to be reared outdoors, in the belief that this made them stronger laying birds, better able to withstand life in the cage. Now they have been moved inside, and in many cases are placed in cages almost from birth, since with tiers of cages more birds can be accommodated in each shed and the overhead per bird is correspondingly lower. Since the birds grow rapidly, however, they have to be moved to larger cages and this is a disadvantage, since "mortality may be a little higher.... Broken legs and bruised heads are bound to occur when you move birds."[31]

Whatever the method of rearing used, all the big egg producers now keep their laying hens in cages. (These are often referred to as "batteries" or "battery cages," not because there is anything electrical about them, but from the original meaning of the word "battery" as "a set of similar or connected units of equipment.") When cages were first introduced there was only one bird to a cage, the idea being that the farmer could then tell which birds were not laying enough eggs to give an economic return on their food. Those birds would then be killed. Then it was found that more birds could be housed and costs per bird reduced if two birds were put in each cage. That was only the first step. Now there is no question of keeping a tally of each bird's eggs. Cages are used because of the greater number of birds who can be housed, warmed, fed, and watered in one building, and the greater use that can be made of labor-saving automatic equipment.

The economic demand that labor costs be kept to an absolute minimum means that laying hens get no more individual attention than broilers. Alan Hainsworth, owner of a poultry farm in upstate New York, told an inquiring local reporter that four hours a day was all he needed for the care of his 36,000 laying hens, while his wife looked after the 20,000 pullets: "It takes her about 15 minutes a day. All she checks is their automatic feeders, water cups and any deaths during the night."

This kind of care does not ensure a happy flock, though, as the reporter's description shows:

> Walk into the pullet house and the reaction is immediate—complete pandemonium. The squawking is loud and intense as some 20,000 birds shove to the farthest side of their cages in fear of the human intruders.[32]

Julius Goldman's Egg City, fifty miles northwest of Los Angeles, was one of the first million-plus layer units. Already in 1970, when the *National Geographic Magazine* did an enthusiastic survey of what were then still relatively novel farming methods, it consisted of two million hens divided into block-long buildings containing 90,000 hens each, five birds to a sixteen-by-eighteen-inch cage. Ben Shames, Egg City's executive vice-president, explained to their reporter the methods used to look after so many birds:

> We keep track of the food eaten and the eggs collected in 2 rows of cages among the 110 rows in each building. When production drops to the uneconomic point, all 90,000 birds are sold to processors for potpies or chicken soup. It doesn't pay to keep track of every row in the house, let alone individual hens; with 2 million birds on hand you have to rely on statistical samplings.[33]

In most egg factories the cages are stacked in tiers, with food and water troughs running along the rows filled automatically from a central supply. The cages have sloping wire floors. The slope—usually a gradient of one in five—makes it more difficult for the birds to stand comfortably, but it causes the eggs to roll to the front of the cage where they can easily be collected by hand or, in the more modern plants, carried by conveyor belt to a packing plant.

The wire floor also has an economic justification. The excrement drops through and can be allowed to pile up for many months until it is all removed in a single operation. (Some producers remove it more frequently; others don't.) Unfortunately the claws of the hen are not well adapted to living on wire, and reports of damage to hens' feet are common whenever anyone bothers to make an examination. Without any solid ground to wear them down, the birds' toenails become very long and may get permanently entangled in the wire. A former president of a national poultry organization reminisced in an industry magazine about this:

> We have discovered chickens literally grown fast to the cages. It seems that the chickens' toes got caught in the wire mesh in some manner and would not loosen. So, in time, the flesh of the toes grew completely around the wire. Fortunately for the birds, they were caught near the front of the cages where food and water were easily available to them.[34]

Next we must consider the amount of living space available to laying hens in cages. In Britain, the Protection of Birds Act, passed in 1954, is intended to prevent cruelty to birds. Clause 8, subsection 1 of this law runs as follows:

> If any person keeps or confines any bird whatsoever in any cage or other receptacle which is not sufficient in height, length or breadth to permit the bird to stretch its wings freely, he shall be guilty of an offence against the Act and be liable to a special penalty.

While any caging is objectionable, the principle that a cage should be large enough to allow birds to stretch their wings freely seems an absolute minimum necessary to protect them from an intolerable degree of confinement that frustrates a very basic urge. So may we assume that poultry cages in Britain must at least be large enough to give the birds this minimal freedom? No. The subsection quoted above has a short but significant proviso attached to it:

> Provided that this subsection shall not apply to poultry...

This amazing proviso testifies to the relative strength of desires that emanate from the stomach and those that are based on compassion in a country that has a reputation for kindness to animals. Nothing in the nature of those birds we call "poultry" makes them less desirous of stretching their wings than other birds. The only conclusion we can draw is that the members of the British Parliament are against cruelty except when it produces their breakfast.

There is a close parallel to this in the United States. Under the Animal Welfare Act of 1970 and subsequent revisions, standards have been set requiring cages for animals to "provide sufficient space to allow each animal to make normal postural and social adjustments with adequate freedom of movement." This act applies to zoos, circuses, wholesale pet dealers, and laboratories, but not to animals being reared for food.[35]

So how do cages for laying hens measure up by the minimal standard set for birds in general? To answer this question we need to know that the wingspan of the most common type of hen averages around thirty inches. Cage sizes vary, but according to *Poultry Tribune,*

> a typical size is 12 by 20 inches in which anywhere from one to five layers are housed. Space available per bird varies from 240 to 48 square inches depending on the number of birds per cage. There is a tendency to crowd the layers to reduce building and equipment costs per bird.[36]

Obviously this size is too small for even one bird to stretch her wings fully, let alone five birds in the same cage—and as the last line of the quoted passage hints, four or five birds, not one or two, is the industry standard.

Since the first edition of this book was published, the conditions under which hens are housed in modern intensive farming have been the subject of numerous studies, both by scientific and governmental committees. In 1981 the British House of Commons Agriculture Committee issued a report on animal welfare in which it said "we have seen for ourselves battery cages, both experimental and commercial, and we greatly dislike what we saw." The committee recommended that the British government should take the initiative in having battery cages phased out within five years.[37] Still more telling, however, was a study conducted at the Houghton Poultry Research Station in Britain on the space required by hens for various activities. This study found that the typical hen at rest physically occupies an area of 637 square centimeters, but if a bird is to be able to turn around at ease, she would need a space of 1,681 square centimeters if kept in a single cage. In a five-bird cage, the study concluded that the size of the cage should allow room at the front for all birds, and therefore needed to be not less than 106.5 centimeters long and 41 centimeters deep, giving each bird 873 square centimeters (approximately 42 by 16 inches).[38] The 48 square inches noted above in the *Poultry Tribune* article, when five birds are in the standard twelve-by-twenty-inch cages, converts to just 300 square centimeters. With only four birds in such cages, each bird has 375 square centimeters.

Although the British government has taken no action on the recommendation to take the intitiative in phasing out cages, change is possible. In 1981 Switzerland began a ten-year phase-out of battery cages. By 1987 birds in cages had to have a minimum of 500 square centimeters; and on the first day of 1992, traditional cages will be outlawed and all laying hens will have access to protected, soft-floored nesting boxes.[39] In the Netherlands, conventional battery cages will become illegal in 1994, and hens will have a minimum space allowance of 1,000 square centimeters, as well as access to nesting and scratching areas. More far-reaching still, however, is a Swedish law passed in July 1988 that requires the abolition of cages for hens over the next ten years and states that cows, pigs, and animals raised for their furs must be kept "in as natural an environment as possible."[40]

The rest of Europe is still debating the future of the battery cage. In 1986 the ministers of agriculture of the European Community countries set the minimum space allowance for laying hens at 450 square centimeters. Now it has been decided that this minimum will not become a legal requirement until 1995. Dr. Mandy Hill, deputy director of the British Ministry of Agriculture's Gleadthorpe experimental farm, has estimated that 6.5 million birds in Britain will need to be rehoused, indicating that this many birds at present have less than this ridiculously low minimum.[41] But since the total British laying flock is around 50 million, and approximately 90 percent of these are kept in cages, this also shows that the new minimum will do no more than write into the law the very high stocking densities that most egg producers are already using. Only a minority who squeeze their birds even more tightly than is standard in the industry will have to change. Meanwhile in 1987 the European Parliament recommended that battery cages be phased out in the European Community within ten years.[42] But the European Parliament only has advisory powers, and Europeans anxious to see the end of the cages have nothing to celebrate yet.

The United States, however, lags far behind Europe in even beginning to tackle this problem. The European Community minimum standard of 450 square centimeters is equivalent to seventy square inches per hen; in the United States, United Egg Producers has recommended forty-eight square inches as a U.S. standard.[43] But the space allowed to birds on farms is often still less. At the Hainsworth farm in Mt. Morris, New York, four hens were squeezed into cages twelve inches by twelve inches—36 square inches per bird—and the reporter added: "Some hold five birds when Hainsworth has more birds than room."[44] The truth is that whatever official or semiofficial recommendations there may be, one never knows how many hens are packed into cages unless one goes and looks. In Australia, where a government "Code of Practice" suggests that there should be no more than four hens in an eighteen-by-eighteen-inch cage, an unannounced visit to one farm in the state of Victoria in 1988 revealed seven birds in one cage that size, and five or six in many others. Yet the Department of Agriculture in the state of Victoria refused to prosecute the producer.[45] Seven birds in a cage eighteen inches square have just 289 square centimeters, or forty-six square inches. At these stocking rates a single sheet of typing paper represents the living space for two hens, and the birds are virtually sitting on top of each other.

Under the conditions standard on modern egg farms in the United States, Britain, and almost every other developed nation except, shortly, Switzerland, the Netherlands, and Sweden, every natural instinct the birds have is frustrated. They cannot walk around, scratch the ground, bathe in the dust, build nests, or stretch their wings. They are not part of a flock. They cannot keep out of each other's way, and weaker birds have no escape from the attacks of stronger ones, already maddened by the unnatural conditions. The extraordinary degree of crowding results in a condition that scientists call "stress," resembling the stress that occurs in human beings subject to extreme crowding, confinement, and frustration of basic activities. We saw that in broilers this stress leads to aggressive pecking and cannibalism. In layers, kept for longer periods, the Texas naturalist Roy Bedichek observed other signs:

> I have looked attentively at chickens raised in this fashion and to me they seem to be unhappy.... The battery chickens I have observed seem to lose their minds about the time they would normally be weaned by their mothers and off in the weeds chasing grasshoppers on their own account. Yes, literally, actually, the battery becomes a gallinaceous madhouse.[46]

Noise is another indication of distress. Hens scratching in a field are generally quiet, making only an occasional cluck. Caged hens tend to be very noisy. I have already quoted the reporter who visited the pullet house on the Hainsworth farm and found "complete pandemonium." Here is the same reporter's account of the laying house:

> The birds in the laying house are hysterical. The uproar of the pullet house was no preparation for this. Birds squawk, cackle and cluck as they scramble over one another for a peck at the automatically controlled grain trough or a drink of water. This is how the hens spend their short life of ceaseless production.[47]

The impossibility of building a nest and laying an egg in it is another source of distress for the hen. Konrad Lorenz has described the laying process as the worst torture to which a battery hen is exposed:

> For the person who knows something about animals it is truly heart-rending to watch how a chicken tries again and again to crawl beneath her fellow-cagemates, to search there in vain for cover. Under these circumstances hens will undoubtedly hold back their eggs for as long as possible. Their instinctive reluctance to lay eggs amidst the crowd of their

cagemates is certainly as great as the one of civilised people to defecate in an analogous situation.[48]

Lorenz's view has been supported by a study in which hens were able to gain access to a nesting box only by overcoming increasingly difficult obstacles. Their high motivation to lay in a nest was shown by the fact that they worked just as hard to reach the nesting box as they did to reach food after they had been deprived of food for twenty hours.[49] Perhaps one reason why hens have evolved an instinct to lay eggs in privacy is that the vent area becomes red and moist when the egg is laid, and if this is visible to other birds, they may peck at it. If this pecking draws blood, further pecking will result, which can lead to cannibalism.

Hens also provide another kind of evidence that they never lose their nesting instinct. Several of my friends have adopted a few hens who were at the end of their commercial laying period and about to be sent to the slaughterhouse. When these birds are released in a backyard and provided with some straw, they immediately start to build nests—even after more than a year spent in a bare metal cage. In Switzerland, by the end of 1991, the law will require that laying hens have protected, darkened, and soft-floored or litter-lined nesting boxes. Swiss scientists have even investigated what kind of litter hens prefer and found that both caged hens and hens who had been reared on litter preferred oat husks or wheat straw; as soon as they discovered that they had a choice, none laid eggs on wire floors or even on synthetic grass. Significantly the study found that while nearly all the hens reared on litter had left the nesting boxes forty-five minutes after they were admitted to them, the cage-reared birds seemed to be so entranced with their new-found comforts that at the end of this period 87 percent of them were still sitting there![50]

This story is repeated with other basic instincts thwarted by the cage system. Two scientists watched hens who had been kept in cages for the first six months of their lives and found that within the first ten minutes after release, half of the hens had already flapped their wings, an activity that was barely possible in the cages.[51] The same is true of dustbathing—another important instinctive activity that has been shown to be necessary for maintaining feather quality.[52] A farmyard hen will find a suitable area of fine soil and then form a hollow in it, fluffing up the soil into her feathers and then shaking energetically to remove the dust. The need to do this is instinctive, and present even in caged birds. One study found that birds kept on wire floors had "a higher denudation of the belly" and suggested that "the lack of appropriate material for dustbathing may be an important factor, as it is well known that hens perform dustbathing activities directly on the wire floor."[53] Indeed, another researcher found that hens kept on wire actually engage in dustbathing-like behavior—without any dust to fluff into their feathers—more often than birds kept on sand, although for shorter periods of time.[54] The urge to dustbathe is so strong that hens keep trying to do so, despite the wire floors, and rub the feathers off their bellies in the process. Again, if released from the cages, these birds will take up dustbathing with real relish. It is wonderful to see how a dejected, timid, almost featherless hen can, in a relatively short period, recover both her feathers and her natural dignity when put into a suitable environment.

To appreciate the constant and acute frustration of the lives of hens in modern egg factories it is best to watch a cage full of hens for a short period. They seem unable to stand or perch comfortably. Even if one or two birds were content with their positions, so long as other birds in the cage are moving, they must move too. It is like watching three people trying to spend a comfortable night in a single bed—except that the hens are condemned to this fruitless struggle for an entire year rather than a single night. An added irritation is that after a few months in the cages the birds start to lose their feathers, partly from rubbing against the wire, and partly because other birds are constantly pecking at them. The result is that their skin begins to rub against the wire, and it is common to see birds who have been in the cages for some time with few feathers, and skin rubbed bright red and raw, especially around the tail.

As with broilers, feather-pecking is a sign of stress and, as one of the previously quoted studies put it, "the lack of appropriate stimulation from the physical environment."[55] It has been shown that in an enriched environment, with access to perches, litter in which to scratch, and nesting boxes, hens peck less and do less feather damage than when they are kept in conventional cages.[56] Feather-pecking is itself the cause of further injuries, because, as another group of researchers has noted,

> scratches and torn skin, especially on the back . . . are more likely to occur when the skin on the back is no longer protected by feathers. Thus, fear, feather loss and pain may, at times, all be part of the same syndrome.[57]

Finally, in most cages there is one bird—maybe more than one in larger cages—who has lost the will to resist being shoved aside and pushed underfoot by other birds. Perhaps these are the birds who, in a normal farmyard, would be low in the pecking order; but under normal conditions this would not matter so much. In the cage, however, these birds can do nothing but huddle in a corner, usually near the bottom of the sloping floor, where their fellow inmates trample over them as they try to get to the food or water troughs.

Although after all this evidence it might seem otiose to study whether hens prefer cages or outside runs, Dr. Marian Dawkins of the department of zoology at Oxford University has done just that, and her work provides yet more scientific backing for what has already been said. Given a choice, hens familiar with both grassed runs and cages will go to the run. In fact, most of them will prefer a run with no food on it to a cage that does have food in it.[58]

Ultimately the most convincing way a hen can indicate that her conditions are inadequate is by dying. A high rate of mortality will occur only under the most extreme conditions, since the normal life span of a chicken is far longer than the eighteen months to two years that laying hens are allowed to live. Hens, like humans in concentration camps, will cling tenaciously to life under the most miserable conditions. Yet it is commonplace for an egg farm to lose between 10 and 15 percent of its hens in one year, many of them clearly dying of stress from overcrowding and related problems. Here is one example:

> According to the manager of a 50,000 bird egg ranch near Cucamonga, California, five to ten of his hens succumb daily to confinement stress. (That's between two and four thousand per year.) "These birds," he says, "don't die of any disease. They just can't take the stress of crowded living."[59]

A carefully controlled study by members of the department of poultry science at Cornell University confirmed that crowding increases death rates. Over a period of less than a year, mortality among layers housed three to a twelve-by-eighteen-inch cage was 9.6 percent; when four birds were put in the same cage, mortality jumped to 16.4 percent; with five birds in the cage, 23 percent died. Despite these findings, the researchers advised that "under most conditions Leghorn layers should be housed at four birds per 12 by 18 inch cage," since the greater total number of eggs obtained made for a larger return on capital and labor, which more than compensated for the higher costs in respect of what the researchers termed "bird depreciation."[60] Indeed, if egg prices are high, the report concluded, "five layers per cage make a greater profit." This situation parallels that which we have already seen demonstrated with regard to broilers, and again proves that animal factory managers can make bigger profits by keeping their animals in more crowded conditions, even though more of the animals may die under those conditions. Since laying eggs is a bodily function (like ovulation for a woman) hens continue to lay eggs, even when they are kept in conditions that frustrate all their behavioral needs.

So the hens that produce our eggs live and die. Perhaps those who die early are the lucky ones, since their hardier companions have nothing in store for them except another few months of crowded discomfort. They lay until their productivity declines, and then they are sent off to be slaughtered and made into chicken pies or soups, which by then is all they can be used for.

There is only one likely alternative to this routine, and it is not a pleasant one. When egg production begins to drop off it is possible to restore the hens' reproductive powers by a procedure know as "force-molting." The object of force-molting is to make the hen go through the physiological processes associated, under natural conditions, with the seasonal loss of old plumage and growth of fresh feathers. After a molt, whether natural or artificial, the hen lays eggs more frequently. To induce a hen to molt when she is living in a controlled-environment shed without seasonal changes in temperature or length of light requires a considerable shock to her system. Typically the hens will find that their food and water, which have been freely available to them until this time, are suddenly cut off. For instance, until quite recently a British Ministry of Agriculture booklet advised that the second day of a forced molt should be as follows:

> No food, light or water. Make sure the food troughs are really empty, clean out any remaining mash, collect eggs, then turn off the water and lights and leave the birds for 24 hours.[61]

The standard practice was then that after two days water would be restored and food after another day. Over the next few weeks the lighting would be returned to normal and those hens who had survived—some succumbed from the shock—might be expected to be sufficiently productive to be worth keeping for another six months or so. Since 1987, as a result of pressure from animal welfare groups, this method of force-molting has been illegal in Britain, and hens must get food and water every day. In the United States it is still entirely legal. Many poultry farmers, however, do not consider this procedure worth the trouble; hens are cheap, so they prefer to get a new flock as soon as the present one is past its peak.

To the very end, egg producers allow no sentiment to affect their attitudes to the birds who have laid so many eggs for them. Unlike the murderer who gets a special meal before being hanged, the condemned hens may get no food at all. "Take feed away from spent hens" advises a headline in *Poultry Tribune*, and the article below tells farmers that food given to hens in the thirty hours prior to slaughter is wasted, since processors pay no more for food that remains in the digestive tract.[62]

Pigs

Of all the animals commonly eaten in the Western world, the pig is without doubt the most intelligent. The natural intelligence of a pig is comparable and perhaps even superior to that of a dog; it is possible to rear pigs as companions to human beings and train them to respond to simple commands much as a dog would. When George Orwell put the pigs in charge in *Animal Farm* his choice was defensible on scientific as well as literary grounds.

The high intelligence of pigs must be borne in mind when we consider whether the conditions in which they are reared are satisfactory. While any sentient being, intelligent or not, should be given equal consideration, animals of different capacities have different requirements. Common to all is a need for physical comfort. We have seen that this elementary requirement is denied to hens; and, as we shall see, it is denied to pigs as well. In addition to physical comfort, a hen requires the structured social setting of a normal flock; she may also miss the warmth and reassuring clucks of the mother hen immediately after hatching; and research has provided evidence that even a chicken can suffer from simple boredom.[63] To whatever extent this is true of chickens, it is certainly true, and to a greater extent, of pigs. Researchers at Edinburgh University have studied commercial pigs released into a seminatural enclosure, and have found that they have consistent patterns of behavior: they form stable social groups, they build communal nests, they use dunging areas well away from the nest, and they are active, spending much of the day rooting around the edge of the woodlands. When sows are ready to give birth, they leave the communal nest and build their own nest, finding a suitable site, scraping a hole, and lining it with grass and twigs. There they give birth and live for about nine days, until they and their piglets rejoin the group.[64] As we shall see, factory farming makes it impossible for the pigs to follow these instinctive behavior patterns.

Pigs in modern factory farms have nothing to do but eat, sleep, stand up, and lie down. Usually they have no straw or other bedding material, because this complicates the task of cleaning. Pigs kept in this way can hardly fail to put on weight, but they will be bored and unhappy. Occasionally farmers notice that their pigs like stimulation. One British farmer wrote to *Farmer's Weekly* describing how he had housed pigs in a derelict farmhouse and found that they played all around the building, chasing each other up and down the stairs. He concluded:

> Our stock need variety of surroundings.... Gadgets of different make, shape and size should be provided.... Like human beings, they dislike monotony and boredom.[65]

This common-sense observation has now been backed up by scientific studies. French research has shown that when deprived or frustrated pigs are provided with leather strips or chains to pull, they have reduced levels of corticosteroids (a hormone associated with stress) in their blood.[66] British research has shown that pigs kept in a barren environment are so bored that if they are given both food and an earth-filled trough, they will root around in the earth before eating.[67]

When kept in barren, overcrowded conditions pigs are prone to "vice," as hens are. Instead of feather-pecking and cannibalism pigs take to biting each other's tails. This leads to fighting in the pig pen and reduces gains in weight. Since pigs do not have beaks, farmers cannot debeak them to prevent this, but they have found another way of eliminating the symptoms without altering the conditions that cause the trouble: they cut off the pigs' tails.

According to the U.S. Department of Agriculture:

> Tail docking has become a common practice to prevent tail biting of pigs in confinement. It should be done by all producers of feeder pigs. Cut tails 1/4 to 1/2 inch from the body with side-cutting pliers or another blunt instrument. The crushing action helps to stop bleeding. Some producers use a chicken debeaker for docking; this also cauterizes the cut surface.[68]

This is a doubly disgraceful recommendation. But before I explain why, here are the candid views of a pig producer on tail docking:

> They hate it! The pigs just hate it! And I suppose we could probably do without tail-docking if we gave them more room, because they don't get so crazy and mean when they have more space. With enough room, they're actually quite nice animals. But we can't afford it. These buildings cost a lot.[69]

In addition to more space, another possible remedy is suggested by a leading farm animal researcher:

> The probable underlying cause . . . is that pigs are using species-typical activities in an unusual way because no suitable object is available. The lower incidence of tail-biting in units with straw bedding is probably due, at least in part, to the "recreational" effects of the straw.[70]

Now we can see why the U.S. Department of Agriculture's cold recommendations are disgraceful. First, no suggestion is made that pigs having their tails cut off should be given pain-killers or anesthetics. Second, there is no mention of the fact that the need to deprive pigs of their tails is an indication that the pigs are too crowded or deprived of straw or anything else to attract their interest. The problem seems to be that bored pigs gnaw at any attractive object, and if gnawing on the tail of another pig should produce an injury and draw blood, some pigs will be attracted to the blood and begin biting in earnest.[71] It is entirely typical of the mentality of modern animal production, though, that the answer of both the Department of Agriculture and of the pig producers is to mutilate the animals instead of giving them the living conditions they need.

Another respect in which confined pigs resemble confined hens is that they suffer from stress, and in many cases they die from it. Because in pig farming an individual pig makes a much larger contribution to total profits than the individual hen does, the pig farmer has to take this problem more seriously than the poultry farmer does. There is a name for the condition—the "porcine stress syndrome"—and the symptoms have been described in one farming journal as; "extreme stress...rigidity, blotchy skin, panting, anxiety, and often—sudden death."[72] The condition is especially upsetting to producers because, as the same article says: "Painfully, you often lose PSS hogs when they near market weight, with a full investment of feed."

There is also strong evidence that porcine stress syndrome has increased dramatically as confinement rearing has become more common.[73] Confined pigs are so delicate that any disturbance can bring on the symptoms, including a strange noise, sudden bright lights, or the farmer's dog. Nevertheless, if one were to suggest reducing stress by eliminating confinement methods of production, the reaction would almost certainly be that expressed in *Farmer and Stockbreeder* some years ago, when confinement was still fairly new and stress-related deaths were just beginning to be noticed:

> These deaths in no way nullify the extra return obtained from the higher total output.[74]

In the pig industry, in contrast to the broiler and egg industry, total confinement is not yet universal. But the trend is in that direction. A University of Missouri survey revealed that as long ago as 1979, 54 percent of all medium-sized producers and 63 percent of all large producers had total confinement facilities.[75] Increasingly, it is the large producers that dominate the industry. In 1987 William Haw, president of National Farms, Inc., said that "within ten years the hog business will be the same as the broiler chicken industry is now, with fewer than 100 operators of any significance."[76] It is the old story: small family farms are being pushed out of business by large factories, each "manufacturing" between 50,000 and 300,000 pigs a year. Tyson Foods, the largest broiler company in the world, slaughtering more that 8.5 million birds a week, has now entered the pig market. The company runs sixty-nine farrowing and nursery complexes and sends to slaughter more than 600,000 pigs per year.[77]

So most pigs now spend their entire lives indoors. They are born and suckled in a farrowing unit, raised initially in a nursery, and brought to slaughter weight in a growing-feeding unit. Unless they are to be used as breeders, they are sent to market at between five and six months of age weighing about 220 pounds.

The desire to cut labor costs has been one major reason for the shift to confinement. With an intensive system, one man is said to be able to handle the entire operation, thanks to automated feeding and slatted floors that allow the manure to drop through for easy disposal. Another saving, with this as with all other confinement systems, is that with less room to move about, the pig will burn up less of its food in "useless" exercise, and so can be expected to put on more weight for each pound of food consumed. In all of this, as one pig producer said, "What we are really trying to do is modify the animal's environment for maximum profit."[78]

In addition to stress, boredom, and crowding, modern pig confinement units create physical problems for the pigs. One is the atmosphere. Here is a quotation from the herdsman at Lehman Farms, of Strawn, Illinois:

> The ammonia really chews up the animals' lungs.... The bad air's a problem. After I've been working in here awhile, I can feel it in my own lungs. But at least I get out of here at

> night. The pigs don't, so we have to keep them on tetracycline, which really does help control the problem.[79]

Nor is this a producer of particularly low standards. The year before this statement was made, Lehman had been named Illinois Pork All-American by the National Pork Producers' Council.

Another physical problem for the pigs is that the floors of confinement units are designed for ease of maintenance and the elimination of chores like manure disposal, rather than for the comfort of the animals. In most units the floors are either slatted or solid concrete. Neither is satisfactory; both damage the feet and legs of the pigs. Studies have shown extraordinarily high rates of foot damage, but a discussion of slatted floors by the editor of *Farmer and Stockbreeder* stated the producers' attitude to this question clearly:

> The commonsense approach at this stage in our knowledge is that for expendable stock the slatted floor seems to have more merit than disadvantage. The animal will usually be slaughtered before serious deformity sets in. On the other hand, breeding stock, with a longer working life before it, must grow and keep good legs; risk of damage here would seem to outweigh the advantages.[80]

An American producer put it more tersely:

> We don't get paid for producing animals with good posture around here. We get paid by the pound.[81]

While the fact that the animal will usually be slaughtered before serious deformity sets in may minimize the financial loss to the producer, it can hardly comfort the animals, standing continuously on unsuitable flooring, acquiring foot or leg deformities that would become serious were the pigs not to be slaughtered at an early age.

The solution, of course, is to take the pigs off bare concrete floors. One British pig farmer with three hundred sows did just that, putting his pigs in outdoor, straw-lined yards with kennels. He reported:

> During the time that all our pregnant sows were close housed, we suffered considerable losses due to abrasion, twisted gut, lameness, sores and hip problems.... We can demonstrate that we have few lame sows and minimal damage due to fighting in the [outdoor] group.[82]

Very few pigs have the luxury of straw yards, and the overall trend is still in the wrong direction. Taking a lead again from the poultry industry, pig farmers in Holland, Belgium, and England have begun rearing baby pigs in cages. American producers are now trying it. Apart from the usual desire for faster gains on less feed and more tender meat because of restricted opportunities to exercise, the main advantage of cages is that the piglets can be weaned from their mothers earlier. This means that the sow's lactation will cease, and within a few days she will become fertile. She will then be made pregnant again, either by a boar or by artificial insemination. The result is that with early weaning a sow can produce an average of 2.6 litters a year, instead of the maximum of 2.0 that can be produced if the pigs are allowed to suckle for three months as they would naturally do.[83]

Most cage pig rearers allow the piglets to nurse from their mothers for at least a week before moving them to cages; but Dr. J. Frank Hurnick, a Canadian agricultural researcher, has recently developed a mechanical sow. According to one report, "Hurnick's success could allow intensive breeding efforts to be directed at increasing litter size. So far, litters have always been limited by the capacity of the sow mammary system."[84] By the combination of mechanical nursing and other novel techniques like superovulation, which increases the number of fertile eggs the sow produces, researchers foresee highly automated systems of pig production producing as many as forty-five pigs per sow per year, instead of the sixteen that have been the average.

Two aspects of these developments are alarming. First there is the effect on the baby pigs, deprived of their mothers and confined in wire cages. In mammals, the early separation of mother and child causes distress to both. As for the cages themselves, an ordinary citizen who kept dogs in similar conditions for their entire lives would risk prosecution for cruelty. A pig producer who keeps an animal of comparable intelligence in this manner, however, is more likely to be rewarded with a tax concession or, in some countries, a direct government subsidy.

The second alarming aspect of the new techniques is that the sow is being turned into a living reproduction machine. "The breeding sow should be thought of, and treated as, a valuable piece of machinery whose function is to pump out baby pigs like a sausage machine."[85] So said a leading corporate manager with Wall's Meat Company; and the United States Department of Agriculture actually encourages producers to think of the pig in this way: "If the sow is considered a pig manufacturing unit, then improved management at farrowing and on through weaning will result in more pigs weaned per sow per year."[86] Under the best conditions there is little joy in an existence that consists of pregnancy, birth, having one's babies taken away, and becoming pregnant again so that the cycle can be repeated—and sows do not live under the best conditions. They are closely confined for both pregnancy and birth. While pregnant they are usually locked into individual metal stalls two feet wide and six feet long, or scarcely bigger than the sow herself; or they may be chained by a collar around the neck; or they may be in stalls yet still be chained. There they will live for two or three months. During all that time, they will be unable to walk more than a single step forward or backward, or to turn around, or to exercise in any other way. Again, savings on feed and labor are the reason for this brutal form of solitary imprisonment.

When the sow is ready to give birth she is moved—but only to a "farrowing pen." (Humans give birth, but pigs "farrow.") Here the sow may be even more tightly restricted in her movements than she was in her stall. A device nicknamed "the iron maiden," consisting of an iron frame that prevents free movement, has been introduced and widely used in many countries. The ostensible purpose is to stop the sow rolling onto and crushing her piglets, but this could also be achieved by providing her with more natural conditions.

When the sow is confined both while pregnant and while nursing—or when she is deprived of the opportunity to nurse—she is tightly restricted for almost the whole of her life. In confinement, the environment is monotonous and the pig has scant chance to choose or alter her environment. The U.S. Department of Agriculture admits that "the sow kept in a crate cannot fulfill her strong instinct to build a nest" and this frustration can contribute to farrowing and lactation problems.[87]

The sows themselves make it clear what they think of this form of confinement. At the University of Wageningen, in the Netherlands, G. Cronin obtained a Ph.D. for a study of the behavior of confined sows. Here is his description of how they behave when first put in a stall with a tether:

> The sows threw themselves violently backwards, straining against the tether. Sows thrashed their heads about as they twisted and turned in their struggle to free themselves. Often loud screams were emitted and occasionally individuals crashed bodily against the side boards of the tether stalls. This sometimes resulted in sows collapsing to the floor.[88]

These violent attempts to escape can last up to three hours. When they subside, Cronin reports, the sows lie still for long periods, often with their snouts thrust under the bars, making occasional quiet groans and whining noises. After a further period, the sows show other signs of stress, such as gnawing the bars of their stalls, chewing when there is nothing to chew, waving their heads back and forth, and so on. This is known as stereotypical behavior. Anyone who has been to a zoo that keeps lions, tigers, or bears in barren concrete enclosures will have seen stereotypical behavior—the animals pace endlessly up and down the fences of their cages. The sow does not have even this opportunity. As we have seen, in natural conditions the sow is a highly active animal, spending several hours a day finding food, eating, and exploring her environment. Now, gnawing at the bars of the stall is, as a veterinarian has noted, "one of the few physical expressions available to her in her barren environment."[89]

In 1986 the Scottish Farm Buildings Investigation Unit, a government-supported research organization, published a review of the scientific evidence on the question: "Does close confinement cause distress in sows?" After discussing more than twenty different studies, the report likened the stereotypical behavior of sows to obsessive-compulsive behavior in neurotic human beings who continuously wash or wring their hands. Its answer to the question investigated was unequivocal: "The close confinement of sows causes severe distress."[90] The British Farm Animal Welfare Council, an official advisory body to the British government, reached the same conclusion, in more official language, in its 1988 report:

> Both stall and tether systems fail to meet certain welfare criteria to which we attach particular importance. As a result of their design the animals housed in them are prevented from exercising and from displaying most natural behaviour patterns; in the wide range of systems seen by members there was little scope to reduce the continuing stress which can be caused by confinement in these systems.... We recommend...that the Government should introduce legislation as a matter of urgency to prevent all further installations of units of these designs.[91]

Only when the sow is placed with the boar does she have a short period of freedom in a larger pen—although this is still likely to be indoors. For at least ten months in every year, the pregnant and nursing sow will be unable to walk around. When artificial insemination is more widely used, this sensitive animal will be denied her last chance to exercise, as well as the only remaining natural contact she has with another member of her species, apart for her fleeting contact with her offspring.

In 1988, after more than twenty years of confinement of sows, an important study was published showing that the unfortunate confined sows and boars used for breeding have yet another source of distress: they are kept permanently hungry. Animals being fattened for market are given as much as they will eat; but to give breeding animals more than the bare minimum required to keep them reproducing is, from the producer's point of view, simply a waste of money. The study showed that pigs fed the rations recommended by the Agricultural Research Council in Great Britain are getting only 60 percent of what they would eat if they had more food available. Moreover, their readiness to press levers in order to get additional food was much the same after eating their daily rations as it was before, indicating that they were still hungry immediately after feeding. As the scientists concluded:

> Commercial levels of feeding for pregnant sows and boars, whilst meeting the needs of the producer, do not satisfy feeding motivation. It has often been assumed that high production levels cannot be achieved in the absence of adequate welfare. Yet the hunger resulting from the low food levels offered to the pig breeding population may act as a major source of stress.[92]

Once again, the producer's profits and the interests of the animal are in conflict. It is truly amazing how often this can be demonstrated—while the agribusiness lobby constantly assures us that only happy, well-cared-for animals can be productive.

Of all the forms of intensive farming now practiced, the veal industry ranks as the most morally repugnant. The essence of veal raising is the feeding of a high-protein food to confined, anemic calves in a manner that will produce a tender, pale-colored flesh that will be served to the patrons of expensive restaurants. Fortunately this industry does not compare in size with poultry, beef, or pig production; nevertheless it is worth our attention because it represents an extreme, both in the degree of exploitation to which it subjects the animals and in its absurd inefficiency as a method of providing people with nourishment.

Veal is the flesh of a young calf. The term was originally reserved for calves killed before they had been weaned from their mothers. The flesh of these very young animals was paler and more tender than that of a calf who had begun to eat grass; but there was not much of it, since calves begin to eat grass when they are a few weeks old and still very small. The small amount available came from the unwanted male calves produced by the dairy industry. A day or two after being born they were trucked to market where, hungry and frightened by the strange surroundings and the absence of their mothers, they were sold for immediate delivery to the slaughterhouse.

Then in the 1950s veal producers in Holland found a way to keep the calf alive longer without the flesh becoming red or less tender. The trick depends on keeping the calf in highly unnatural conditions. If calves were left to grow up outside they would romp around the fields, developing muscles that would toughen their flesh and burning up calories that the producer must replace with costly feed. At the same time they would eat grass, and their flesh would lose the pale color that the flesh of newborn calves has. So the specialist veal producers take their calves straight from the auction ring to a confinement unit. Here, in a converted barn or specially built shed, they have rows of wooden stalls, each 1 foot 10 inches wide by 4 feet 6 inches long. It has a slatted wooden floor, raised above the concrete floor of the shed. The calves are tethered by a chain around the neck to prevent them from turning in their stalls when they are small. (The chain may be removed when the calves grow too big to turn around in such narrow stalls.) The stall has no straw or other bedding, since the calves might eat it, spoiling the paleness of their flesh. They leave their stalls only to be taken out to slaughter. They are fed a totally liquid diet, based on nonfat milk powder with vitamins, minerals, and growth-promoting drugs added. Thus the calves live for the next sixteen weeks. The beauty of the system, from the producers' point of view, is that at this age the veal calf may weigh as much as four hundred pounds, instead of the ninety-odd pounds that newborn calves weigh; and since veal fetches a premium price, rearing veal calves in this manner is a profitable occupation.

This method of raising calves was introduced to the United States in 1962 by Provimi, Inc., a feed manufacturer based in Watertown, Wisconsin. Its name comes from the "proteins, vitamins, and minerals" of which its feeds are composed—ingredients that, one might think, could be put to better use than veal raising. Provimi, according to its own boast, created this "new and complete concept in veal raising" and it is still by far the largest company in the business, controlling 50 to 75 percent of the domestic market. Its interest in promoting veal production lies in developing a market for its feed. Describing what it considered "optimum veal production," Provimi's now defunct newssheet, *The Stall Street Journal*, gives us an insight into the nature of the industry, which in the United States and some European countries has remained essentially unchanged since its introduction:

> The dual aims of veal production are firstly, to produce a calf of the greatest weight in the shortest possible time and

> secondly, to keep its meat as light colored as possible to fulfill the consumer's requirement. All at a profit commensurate to the risk and investment involved.[93]

The narrow stalls and their slatted wooden floors are a serious source of discomfort to the calves. When the calves grow larger, they cannot even stand up and lie down without difficulty. As a report from a research group headed by Professor John Webster of the animal husbandry unit at the School of Veterinary Science, University of Bristol, in England, noted:

> Veal calves in crates 750 mm wide cannot, of course, lie flat with their legs extended.... Calves may lie like this when they feel warm and wish to lose heat.... Well-grown veal calves at air temperatures above 20 degrees C [68 degrees F] may be uncomfortably hot. Denying them the opportunity to adopt a position designed to maximise heat loss only makes things worse.... Veal calves in boxes over the age of 10 weeks were unable to adopt a normal sleeping position with their heads tucked into their sides. We conclude that denying veal calves the opportunity to adopt a normal sleeping posture is a significant insult to welfare. To overcome this, the crates would need to be at least 900 mm wide.[94]

American readers should note that 750 millimeters is equivalent to 2 feet 6 inches, and 900 millimeters to 3 feet, both considerably more than standard 1 foot 10 inch crates used in the United States.

The crates are also too narrow to permit the calf to turn around. This is another source of frustration. In addition, a stall too narrow to turn around in is also too narrow to groom comfortably in; and calves have an innate desire to twist their heads around and groom themselves with their tongues. As the University of Bristol researchers said:

> Because veal calves grow so fast and produce so much heat they tend to shed their coats at about 10 weeks of age. During this time they have a great urge to groom themselves. They are also particularly prone to infestation with external

> parasites, especially in mild, humid conditions. Veal calves in crates cannot reach much of their body. We conclude that denying the veal calf the opportunity to groom itself thoroughly is an unacceptable insult to welfare whether this is achieved by constraining its freedom of movement or, worse, by the use of a muzzle.[95]

A slatted wooden floor without any bedding is hard and uncomfortable; it is rough on the calves' knees as they get up and lie down. In addition, animals with hooves are uncomfortable on slatted floors. A slatted floor is like a cattle grid, which cattle always avoid, except that the slats are closer together. The spaces, however, must still be large enough to allow most of the manure to fall or be washed through, and this means that they are large enough to make the calves uncomfortable on them. The Bristol team described the young calves as "for some days insecure and reluctant to change position."

The young calves sorely miss their mothers. They also miss something to suck on. The urge to suck is strong in a baby calf, as it is in a baby human. These calves have no teat to suck on, nor do they have any substitute. From their first day in confinement—which may well be only the third or fourth day of their lives—they drink from a plastic bucket. Attempts have been made to feed calves through artificial teats, but the task of keeping the teats clean and sterile is apparently not worth the producer's trouble. It is common to see calves frantically trying to suck some part of their stalls, although there is usually nothing suitable; and if you offer a veal calf your finger you will find that he immediately begins to suck on it, as human babies suck their thumbs.

Later the calf develops a need to ruminate—that is, to take in roughage and chew the cud. But roughage is strictly forbidden because it contains iron and will darken the flesh, so, again, the calf may resort to vain attempts to chew the sides of his stall. Digestive disorders, including stomach ulcers, are common in veal calves. So is chronic diarrhea. To quote the Bristol study once again:

> The calves are deprived of dry feed. This completely distorts the normal development of the rumen and encourages the

development of hair balls which may also lead to chronic indigestion.[96]

As if this were not enough, the calf is deliberately kept anemic. Provimi's *Stall Street Journal* explains why:

> Color of veal is one of the primary factors involved in obtaining "top-dollar" returns from the fancy veal markets.... "Light color" veal is a premium item much in demand at better clubs, hotels and restaurants. "Light color" or pink veal is partly associated with the amount of iron in the muscle of the calves.[97]

So Provimi's feeds, like those of other manufacturers of veal feeds, are deliberately kept low in iron. A normal calf would obtain iron from grass and other forms of roughage, but since veal calves are not allowed this, they become anemic. Pale pink flesh is in fact anemic flesh. The demand for flesh of this color is a matter of snob appeal. The color does not affect the taste and it certainly does not make the flesh more nourishing—it just means that it lacks iron.

The anemia is, of course, controlled. Without any iron at all the calves would drop dead. With a normal intake their flesh will not fetch as much per pound. So a balance is struck which keeps the flesh pale and the calves—or most of them—on their feet long enough for them to reach market weight. The calves, however, are unhealthy and anemic animals. Kept deliberately short of iron, they develop a craving for it and will lick any iron fittings in their stalls. This explains the use of wooden stalls. As Provimi tells its customers:

> The main reason for using hardwood instead of metal box-stalls is that metal may affect the light veal color.... Keep all iron out of reach of your calves.[98]

And again:

> It is also necessary that calves do not have access to a continuous source of iron. (Water supplied should be checked. If a high level of iron [excess of 0.5 ppm] is present an iron filter

should be considered.) Calf crates should be constructed so calves have no access to rusty metal.[99]

The anemic calf's insatiable craving for iron is one of the reasons the producer is anxious to prevent him turning around in his stall. Although calves, like pigs, normally prefer not to go near their own urine or manure, urine does contain some iron. The desire for iron is strong enough to overcome the natural repugnance, and the anemic calves will lick the slats that are saturated with urine. The producer does not like this, because it gives calves a little iron and because in licking the slats the calves may pick up infections from their manure, which falls on the same spot as their urine.

We have seen that in the view of Provimi, Inc., the twin aims of veal production are producing a calf of the greatest possible weight in the shortest possible time and keeping the meat as light in color as possible. We have seen what is done to achieve the second of these aims, but there is more to be said about the techniques used to achieve fast growth.

To make animals grow quickly they must take in as much food as possible, and they must use up as little of this food as possible in their daily life. To see that the veal calf takes in as much as possible, most calves are given no water. Their only source of liquid is their food—the rich milk replacer based on powdered milk and added fat. Since the buildings in which they are housed are kept warm, the thirsty animals take in more of their food than they would do if they could drink water. A common result of this overeating is that the calves break out in a sweat, rather like, it has been said, an executive who has had too much to eat too quickly.[100] In sweating, the calf loses moisture, which makes him thirsty, so that he overeats again next time. By most standards this process is an unhealthy one, but by the standards of the veal producer aiming at producing the heaviest calf in the shortest possible time, the long-term health of the animal is irrelevant, so long as he survives to be taken to market; and so Provimi advises that sweating is a sign that "the calf is healthy and growing at capacity."[101]

Getting the calf to overeat is half the battle; the other half is ensuring that as much as possible of what has been eaten goes toward putting on weight. Confining the calf so that he cannot exercise is one requirement for achieving this aim. Keeping the barn warm also contributes to it, since a cold calf burns calories just to keep warm. Even warm calves in their stalls are apt to become restless, however, for they have nothing to do all day except at their two mealtimes. A Dutch researcher has written:

> Veal calves suffer from the inability to do something.... The food-intake of a veal calf takes only 20 minutes a day! Besides that there is nothing the animal can do.... One can observe teeth grinding, tail wagging, tongue swaying and other stereotype behavior.... Such stereotype movements can be regarded as a reaction to a lack of occupation.[102]

To reduce the restlessness of their bored calves, many veal producers leave the animals in the dark at all times, except when they are being fed. Since the veal sheds are normally windowless, this simply means turning off the lights. Thus the calves, already missing most of the affection, activity, and stimulation that their natures require, are deprived of visual stimulation and of contact with other calves for more than twenty-two hours out of every twenty-four. Illnesses have been found to be more persistent in dark sheds.[103]

Calves kept in this manner are unhappy and unhealthy animals. Despite the fact that the veal producer selects only the strongest, healthiest calves to begin with, uses a medicated feed as a routine measure, and gives additional injections at the slightest sign of illness, digestive, respiratory, and infectious diseases are widespread. It is common for a veal producer to find that one in ten of a batch of calves do not survive the fifteen weeks of confinement. Between 10 and 15 percent mortality over such a short period would be disastrous for anyone raising calves for beef, but veal producers can tolerate this loss because the high-priced restaurants are prepared to pay well for their products.

Given the cozy relationship that normally exists between veterinarians working with farm animals and intensive producers (it is, after all, the owners, not the animals, who pay the bills), it gives us some indication of the extreme conditions under which veal calves are kept to learn that this is one aspect of animal production that has strained relations between veterinarians and producers. A 1982 issue of *The Vealer* reports:

> Besides waiting too long to call veterinarians for a really sick calf, vets do not look favorable [sic] on relations with veal growers because they have long defied accepted agricultural methods. The feeding of long hay to livestock, in order to maintain a proper digestive system, has been considered a sound practice for years.[104]

The one bright spot in this sorry tale is that the conditions created by the veal crates are so appalling for animal welfare that British government regulations now require that a calf must be able to turn around without difficulty, must be fed a daily diet containing "sufficient iron to maintain it in full health and vigour," and must receive enough fiber to allow normal development of the rumen.[105] These are minimal welfare requirements, and still fall well short of satisfying the needs of calves; but they are violated by almost all the veal units in the United States and by many in Europe.

If the reader will recall that this whole laborious, wasteful, and painful process of veal raising exists for the sole purpose of pandering to people who insist on pale, soft veal, no further comment should be needed.

Dairy

As we have seen, the veal industry is an offshoot of dairying. Producers must ensure that their dairy cows become pregnant every year in order to keep them in milk. Their offspring are taken from them at birth, an experience that is as painful for the mother as it is terrifying for the calf. The mother often makes her feelings plain by constant calling and bellowing for days after her infant is taken. Some female calves will be reared on milk substitutes to become replacements of dairy cows when they reach the age, at around two years, when they can produce milk. Other calves will be sold at between one to two weeks of age to be reared as beef in fattening pens or feedlots. The remainder will be sold to veal producers, who also rely on the dairy industry for the milk diet that is fed to calves to keep them anemic. Even if not sent to a veal unit, as Professor John Webster of the department of animal husbandry at the University of Bristol has written:

> The calf born to the dairy cow is routinely submitted to more insults to normal development than any other farm animal. It is taken from its mother shortly after birth, deprived of its natural food, whole cow's milk, and fed one of a variety of cheaper liquid substitutes.[106]

The dairy cow, once seen peacefully, even idyllically, roaming the hills, is now a carefully monitored, fine-tuned milk machine. The bucolic picture of the dairy cow playing with her calf in the pasture is no part of commercial milk production. Many dairy cows are reared indoors. Some are kept in individual pens with only enough room to stand up and lie down. Their environment is completely controlled: they are fed calculated amounts of feed, temperatures are adjusted to maximize milk yield, and lighting is artificially set. Some farmers have found that a cycle of sixteen hours of light with only eight hours of darkness is conducive to greater output.

After her first calf is taken away, the cow's production cycle begins. She is milked twice, sometimes three times a day, for ten months. After the third month she will be made pregnant again. She will be milked until about six or eight weeks before her next calf is due, and then again as soon as the calf is removed. Usually this intense cycle of pregnancy and hyperlactation can last only about five years, after which the "spent" cow is sent to slaughter to become hamburger or dog food.

In order to obtain the highest output, producers feed cows high-energy concentrates such as soybeans, fish meal, brewing byproducts, and even poultry manure. The cow's peculiar digestive system cannot adequately process this food. The rumen is designed to digest slowly fermenting grass. During peak production, a few weeks after giving birth, the cow often expends more energy than she is able to take in. Because her capacity to produce surpasses her ability to metabolize her feed, the cow begins to break down and use her own body tissues; she begins "milking off her own back."[107]

Dairy cows are sensitive animals who manifest both psychological and physiological disturbances as a result of stress. They have a strong need to identify with their "caretakers." Today's system of dairy production does not allow the farmer more than five minutes a day with each animal. In an article entitled "Dairy

Farms That Don't Need Pastures," one of the largest "milk factories" boasts of an advance that "allows one worker to feed 800 calves in 45 minutes—a job that ordinarily might take several men all day."[108]

Now the rush is on to find ways of interfering with the cow's natural hormonal and reproductive processes to make her produce still more milk. Bovine growth hormone (known in Europe as bovine somatotropin or BST) is being touted as a way of dramatically increasing milk yields. Cows given daily injections of the hormone have been shown to produce about 20 percent more milk. But in addition to the soreness likely to develop from the daily injections, the cows' bodies will be made to work harder still; they will require an even richer diet, and they can be expected to suffer still more from illnesses that already affect dairy cows in large numbers. David Kronfeld, professor of nutrition and chief of large animal medicine at the University of Pennsylvania School of Veterinary Medicine, has said that in one trial over half of the cows given BST were treated for mastitis (a painful inflammation of the mammary gland) compared to none in a control group receiving no BST.[109] Opposition to BST is now coming from dairy farmers as well as animal welfare advocates. This is scarcely surprising, for studies at Cornell University and the United States Congress Office of Technology Assessment have indicated that the adoption of BST by the larger farms could put out of business 80,000 American dairy farmers—half the present number.[110] One dairy farmer from the west of England has pointed out that "the main beneficiaries of these cow injections would be a number of high flying pharmaceutical companies" and has pleaded: "At least let us have milk from contented cows and not from greedy industrialists' pin cushions."[111]

But the production increases brought about by bovine growth hormone are nothing compared to those anticipated by the enthusiasts of new reproductive technology. In 1952 the first calf was produced by means of artificial insemination. Today this is virtually the standard method. In the 1960s the first calves were produced from embryos transferred from one cow to another. This technology means that with the use of hormone injections, a particularly high-yielding cow can be made to produce dozens of eggs at one time. After she is artificially inseminated using semen from a prize bull, the embryos can be flushed out of her womb and transplanted into cheaper surrogate cows through incisions in their flanks. Thus a whole herd can rapidly be bred from only the very best stock. The ability to freeze embryos, developed during the 1970s, has made embryo transfer more easily marketable, and now 150,000 embryo transfers are attempted each year in the United States, with at least 100,000 calves resulting from these attempts. Genetic engineering, and perhaps cloning, will be the next steps in the continuing efforts to create ever more productive animals.[112]

Traditionally, cattle raised for beef in America have roamed freely over the vast open spaces that we see in cowboy movies. But as a supposedly humorous article in the *Peoria Journal Star* indicates, the modern range isn't what it used to be:

> A cowboy's home ain't necessarily on the range. More 'n likely, home is a feedlot where the closest a beef comes to the smell of sage is in a pot roast. This is cowboy'n modern-like. This is Norris Farms where instead of running 700 head on 20,000 acres of sparse-grass prairie, they run 7,000 head on 11 acres of concrete.[113]

By comparison with chickens, pigs, veal calves, and dairy cows, beef cattle still see more of the great outdoors, but the time they have to do so has been diminished. Twenty years ago, cattle would have roamed for about two years. Now, the lucky ones who get to roam at all are rounded up after about six months to be "finished"—that is, to be brought to market weight and condition by being fed a richer diet than grass. For this pupose they are shipped long distances to feedlots. Here for six to eight months they eat corn and other cereals. Then they are sent for slaughter.

The growth of large feedlots has been the dominant trend in the cattle industry. Of the 34 million cattle slaughtered in 1987 in the U.S., 70 percent were sent for slaughter from feedlots. Large feedlots are now responsible for one third of the nation's beef. They are substantial commercial undertakings, often financed by oil companies or Wall Street money looking for tax concessions. Feedlots are profitable because cattle fatten more quickly on grain than on grass. Yet, like dairy cows, beef cattle do not have stomachs suited for the concentrated diet that they receive in feedlots. Often, in an effort to obtain more fiber than their feedlot diets provide, the cattle lick their own and each other's coats. The large amount of hair taken into the rumen may cause abscesses.[114] Diluting the grain with the roughage that cattle need and crave, however, would slow down their weight gain.

Feedlots do not confine cattle as severely as cages confine hens, or stalls confine sows, veal calves, and often dairy cows. Stocking densities have been increasing, but even when they go as high as nine hundred animals to the acre, each animal has fifty square feet of space and can wander around the compound, which may be an acre in area, and not be isolated from other animals. Boredom from the barren, unchanging environment is the problem, not restriction of movement.

One very serious problem is exposure to the elements. In summer the cattle may be out in the sun without shade; in winter they may have no protection from conditions to which they are not naturally suited. During the 1987 snow storms, some farmers reported heavy losses, estimating that as many as 25 to 30 percent of the calves and 5 to 10 percent of the adult cattle were lost. One Colorado farmer reported, "There was little protection for the calves. Most of those calves were lost due to exposure. We got a wet snow, and then it got cold right after that." In another instance, seventy-five calves out of one hundred were lost in the storm.[115]

In Europe some beef producers have followed the lead of the poultry, pig, and veal industries and brought their animals indoors. In the United States, Britain, and Australia, permanent indoor confinement is considered economically unjustified. It protects animals from the weather, but always at the cost of much more crowding, since the cattle producer wants the greatest possible return on the capital invested in the building. Intensively confined beef cattle are generally kept together in groups, in pens rather than in single stalls. Slatted floors are often used for ease of cleaning, although beef cattle, like pigs and veal calves, are uncomfortable on slats and can become lame.

No aspect of animal raising is safe from the inroads of technology and the pressure to intensify production. Baby lambs, those joyous symbols of springtime, have already entered the dark interiors of confinement houses.[116] At the Oregon State University Rabbit Research Center, researchers have developed a cage system for rearing rabbits and are experimenting with stocking densities of two rabbits per square foot.[117] In Australia, selected sheep producing superfine wool have now been brought indoors in individual and group pens—the aim being to keep the fleece clean and long. Wool from these sheep sells for five or six times the usual price.[118] Although the fur trade likes to emphasize its "ranch-raised" furs to minimize the bad publicity it gets from the trapping of wild animals, fur "ranches" are highly intensive. Mink, raccoons, ferrets, and other fur-bearing animals are kept in small wire cages. The beautiful arctic fox, for instance, normally ranges over thousands of acres of tundra: on a fur farm it has a wire cage measuring forty-two by forty-five inches.

We have now surveyed the main trends in animal raising in which traditional methods have been transformed into factory-style animal production. Sadly, as far as the animals are concerned, there has been very little improvement since the first edition of this book was published fifteen years ago. At that time it was already clear that modern production methods are incompatible with any genuine concern for the welfare of the animals. The evidence was first compiled in Ruth Harrison's path-breaking book, *Animal Machines*, published in 1964, and was authoritatively supported by the Brambell committee, a committee appointed by the British minister of agriculture that consisted of the best qualified experts available. In addition to Brambell, himself a noted zoologist, the committee included W. H. Thorpe, the director of the department of animal behavior at Cambridge University, and other experts in veterinary science, animal husbandry, and agriculture. After a thorough investigation, in 1965 they published an eighty-five-page official report. In the report, the committee firmly rejected the argument that productivity is a satisfactory indication of the absence of suffering—the fact that an animal is putting on weight can, they said, be a "pathological condition." They also rejected the view that farm animals don't suffer from confinement because they are bred for it and are used to it. In an important appendix to the report, Thorpe stressed that observations of the behavior of domestic animals have shown that they are "still essentially what they were in the prehistoric wild," with innate behavior patterns and needs that are still present even if the animal has never known natural conditions. Thorpe concluded:

> Certain basic facts are clear enough to justify action. Whilst accepting the need for much restriction, we must draw the line at conditions which completely suppress all or nearly all the natural, instinctive urges and behavior patterns characteristic of actions appropriate to the high degree of social organization as found in the ancestral wild species and which have been little, if at all, bred out in the process of domestication. In particular it is clearly cruel so to restrain an animal for a large part of its life that it cannot use any of its normal locomotary behavior patterns.[120]

Accordingly, the committee's recommendations were based on the following modest but fundamental principle:

> In principle we disapprove of a degree of confinement of an animal which necessarily frustrates most of the major activities which make up its natural behavior. . . . An animal should at least have sufficient freedom of movement to be able without difficulty to turn around, groom itself, get up, lie down and stretch its limbs.[121]

These "five basic freedoms" as they have since been called—to turn around, to groom, to get up, to lie down, and to stretch the limbs freely—are still denied to all caged hens, all sows in stalls and tethers, and all veal calves in crates. Yet since the Brambell committee issued its report a wealth of scientific material has confirmed the verdict of the Brambell committee in all its major aspects. We have already seen, for example, how Thorpe's comments about the retention of natural behavior patterns in domestic animals have been fully borne out by the Edinburgh University study of pigs in a seminatural setting.[122] The fallacy of the argument that animals must be content if they produce is now also universally accepted among scientists. A 1986 study published in *American Scientist* represents an informed view of the argument:

> With respect to domesticated animals, however, this argument can be misleading for several reasons. Farm animals have been selected for their ability to grow and reproduce under a wide range of conditions and circumstances, some adverse. Hens, for example, may continue to lay eggs normally even when severely injured. Furthermore, growth and reproduction are frequently manipulated by practices such as alteration of the photoperiod or the addition of growth-promoting substances like antibiotics to the feed. Finally, on a modern factory farm where a single worker may care for as many as 2,000 head of cattle or 250,000 broiler chickens per year, the practice of measuring growth or reproduction as eggs or pounds of meat produced in relation to construction, fuel, or feed costs provides little information about the productive status of an individual animal.[123]

Dr. Bill Gee, foundation director of the Australian government's Bureau of Animal Health, has said:

> It is claimed that productivity of farm animals is a direct indicator of their welfare. This misconception needs to be buried once and for all. "Welfare" refers to the well-being of individual animals, whereas "productivity" refers to output per dollar spent or per unit of resources.[124]

I have taken care to document the misconception in this argument at several points in this chapter. It would be nice to think that the argument could be buried once and for all, but no doubt it will keep cropping up whenever agribusiness apologists think it useful to lull the consumer into believing that all is well down on the farm.

Some recognition of the weight of evidence against intensive farming methods was given by the European Parliament when in 1987 it considered a report on animal welfare and adopted a policy that contained the following points:

- Putting an end to keeping veal calves in individual crates, and to depriving them of iron and roughage.
- Phasing out the battery cage within ten years.
- Discontinuing keeping sows in individual stalls or on tethers.

- Stopping routine mutilations such as tail docking and the castration of male pigs.[125]

These proposals were passed by a vote of 150 to 0, with 2 abstentions. But as we have already noted, although the European Parliament is made up of elected representatives of all the nations of the European Community, it is an advisory body only. The powerful agribusiness lobby is working hard to stop the policy being put into practice. The resolution stands, nevertheless, as an indication of informed European opinion on these issues. When it comes to actions, not words, a real improvement in the conditions of animals since publication of the first edition of this book has occurred in only a few instances. In Switzerland battery cages for hens are being phased out and eggs from alternative systems of housing hens are already widely available in shops. These new systems allow the birds freedom to walk around, scratch, dustbathe, flutter up to a perch, and lay eggs in protected nest-boxes with suitable nesting materials. Yet the eggs from hens kept in this way are only fractionally more expensive than those from hens kept in cages.[126] In Britain, the only real sign of progress for farm animals is the prohibition of individual stalls for veal calves. It is Sweden that is now showing the way forward in animal welfare, as it has often done in respect of other social reforms; the Swedish laws passed in 1988 will transform conditions for the entire range of farm animals.

Throughout this chapter I have been concentrating on conditions in the United States and Britain. Readers in other countries may be inclined to believe that conditions in their own country are not so bad; but if they live in one of the industrialized nations (other than Sweden) they have no grounds for complacency. In most countries, conditions are much closer to those in the United States than to those recommended above.

Finally, it is important to remember that although the implementation of the Brambell committee's "five freedoms," or of the resolutions of the European Parliament, or even of the new Swedish legislation, would be a major advance in Britain, the United States, and almost anywhere else where factory farming exists, none of these reforms give equal consideration to the similar interests of animals and humans. They represent, to varying degrees, an enlightened and more humane form of speciesism, but speciesism nonetheless. In no country yet has a government body questioned the idea that the interests of animals count less than similar human interests. The issue is always whether there is "avoidable" suffering, and this means suffering that can be avoided while the same animal products are produced, at a cost that is not significantly higher than before. The unchallenged assumption is that humans may use animals for their own purposes, and they may raise and kill them to satisfy their preference for a diet containing animal flesh.

I have concentrated on modern intensive farming methods in this chapter because the general public is largely ignorant of the suffering these methods involve; but it is not only intensive farming that causes animals to suffer. Suffering has been inflicted on animals for human benefit whether they are reared by modern or traditional methods. Some of this suffering has been normal practice for centuries. This may lead us to disregard it, but it is no consolation to the animal on whom it is inflicted. Consider, for example, some of the routine operations to which cattle are still subjected.

Nearly all beef producers dehorn, brand, and castrate their animals. All of these processes can cause severe physical pain. Horns are cut off because horned animals take up more space at a feeding trough or in transit and can harm one another when packed tightly together. Bruised carcasses and damaged hides are costly. The horns are not merely insensitive bone. Arteries and other tissue have to be cut when the horn is removed, and blood spurts out, especially if the calf is not dehorned shortly after birth.

Castration is practiced because steers are thought to put on weight better than bulls—although in fact they seem only to put on more fat—and because of a fear that the male hormones will cause a taint to develop in the flesh. Castrated animals are also easier to handle. Most farmers admit that the operation causes shock and pain to the animal. Anesthetics are generally not used. The procedure is to pin the animal down, take a knife, and slit the scrotum, exposing the testicles. You then grab each testicle in turn and pull on it, breaking the cord that attaches it; on older animals it may be necessary to cut the cord.[127]

Some farmers, to their credit, are troubled by this painful surgery. In an article entitled "The Castration Knife Must Go," C. G. Scruggs, editor of *The Progressive Farmer*, refers to the "extreme stress of castration" and suggests that since lean meat is increasingly in demand, male animals could be left unmutilated.[128] The same view has been expressed in the pig industry, where the practice is similar. According to an article in the British magazine *Pig Farming*:

> Castration itself is a beastly business, even to the hardened commercial pig man. I'm only surprised that the anti-vivisection lobby have not made a determined attack on it.

And since research has now shown a way of detecting the taint that boar meat occasionally has, the article suggests that we "think about giving our castrating knives a rest."[129]

Branding cattle with a hot iron is widely practiced, as a protection against straying and cattle rustlers (who still exist in some parts), as well as to assist record-keeping. Although cattle have thicker skins than humans, their skins are not thick enough to protect them against a red-hot iron applied directly to the skin—the hair having been clipped away first—and held there for five seconds. To permit this operation to be done, the animal is thrown to the ground and pinned down. Alternatively, cattle may be held in a contraption called a "squeeze chute," which is an adjustable crate that can be fitted tightly around the animal. Even so, as one guide notes, "the animal will usually jump when you apply the iron."[130]

As an additional mutilation, cattle are likely to have their ears cut with a sharp knife into special shapes so that, out on the range, they can be identified from a distance or when they are viewed from the front or rear, where the brand would not be visible.[131]

These, then, are some of the standard procedures of traditional methods of rearing cattle. Other animals are treated in similar ways when they are reared for food. And finally, in considering the welfare of animals under traditional systems, it is important to remember that almost all methods involve the separation of mother and young at an early age, and that this causes considerable distress to both. No form of animal raising allows the animals to grow up and become part of a community of animals of varying ages, as they would under natural conditions.

Although castration, branding, and the separation of mother and child have caused suffering to farm animals for centuries, the cruelty of transportation and slaughter aroused the most anguished pleas from the humane movement in the nineteenth century. In the United States the animals were driven from pastures near the Rockies down to the railheads and jammed into railway cars for several days without food until the train reached Chicago. There, in gigantic stockyards reeking of blood and putrefying flesh, those who had survived the journey would wait until their turn came to be dragged and goaded up the ramp at the top of which stood the man with the poleax. If they were lucky, his aim was good; but many were not lucky.

Since that time there have been some changes. In 1906 a federal law was passed limiting the time that animals could spend in a railway car without food or water to twenty-eight hours, or thirty-six hours in special cases. After that time the animals must be unloaded, fed, given water, and rested for at least five hours before the journey is resumed. Obviously, a period of twenty-eight to thirty-six hours in a lurching railway car without food or water is still long enough to cause distress, but it is an improvement. As for slaughter, there has been improvement here too. Most animals are stunned before slaughter now, which means, in theory, that they die painlessly—although as we shall see there are doubts about this, and also important exceptions. Because of these improvements, transportation and slaughter are today lesser problems, I believe, than the factorylike methods of production that turn animals into machines for converting low-priced fodder into high-priced flesh. Nevertheless, any account of what happens to your dinner while it is still an animal would be incomplete without some description of transportation and slaughter methods.

Transportation of animals includes more than the final trip to slaughter. When slaughtering was concentrated at major centers like Chicago this used to be the longest, and in many cases the only, trip the animals made. They grew to market weight on the open ranges on which they were born. When refrigeration techniques enabled slaughtering to become less centralized, the trip to slaughter became correspondingly shorter. Today, however, it is much less common for animals, especially cattle, to be born and raised to market weight in the same region. Young calves may be born in one state—say, Florida—and then trucked to pasture many hundred of miles away—perhaps in west Texas. Cattle who have spent a year out on the ranges in Utah or Wyoming may be rounded up and sent to feedlots in Iowa or Oklahoma. These animals face trips of up to two thousand miles. For them, the journey to the feedlot is likely to be longer and more harrowing than the journey to the slaughterhouse.

The federal law of 1906 provided that animals transported by rail had to be rested, fed, and given water at least every thirty-six hours. It said nothing about animals being transported by truck. Trucks were not used for transporting animals in those days. Over eighty years later, the transportation of animals by truck is still not regulated at the federal level. Repeated attempts have been made to bring the law about trucks into line with that about rail transport, but so far none has succeeded. Accordingly, cattle often spend up to forty-eight or even seventy-two hours inside a truck without being unloaded. Not all truckers would leave cattle this long without rest, food, or water, but some are more concerned with getting the job finished than with delivering their load in good condition.

Animals placed in a truck for the first time in their lives are likely to be frightened, especially if they have been handled hastily and roughly by the men loading the truck. The motion of the truck is also a new experience, and one which may make them ill. After one or two days in the truck without food or water they are desperately thirsty and hungry. Normally cattle eat frequently thoughout the day; their special stomachs require a constant intake of food if the rumen is to function properly. If the journey is in winter, subzero winds can result in severe chill; in summer the heat and sun may add to the dehydration caused by the lack of water. It is difficult for us to imagine what this combination of fear, travel sickness, thirst, near-starvation, exhaustion, and possibly severe chill feels like to the animals. In the case of young calves who may have gone through the stress of weaning and castration only a few days earlier, the effect is still worse. Veterinary experts recommend that, simply in order to improve their prospects of surviving, young calves should be weaned, castrated, and vaccinated at least thirty days prior to being transported. This gives them a chance to recover from one stressful experience before being subjected to another. These recommendations, however, are not always followed.[132]

Although the animals cannot describe their experiences, the reactions of their bodies tell us something. There are two main reactions: "shrinkage" and "shipping fever." All animals lose weight during transportation. Some of this weight loss is due to dehydration and the emptying of the intestinal tract. This loss is easily regained; but more lasting losses are also the rule. For an eight-hundred-pound steer to lose seventy pounds, or 9 percent of his weight, on a single trip is not unusual; and it may take more than three weeks for the animal to recover the loss. This "shrink," as it is known in the trade, is regarded by researchers as an indication of the stress to which the animal has been subjected. Shrink is, of course, a worry to the meat industry, since animals are sold by the pound.

"Shipping fever," a form of pneumonia that strikes cattle after they have been transported, is the other major indicator of stress in transportation. Shipping fever is associated with a virus that healthy cattle have no difficulty in resisting; severe stress, however, weakens their resistance.

Shrinkage and susceptibility to fever are indications that the animals have been subjected to extreme stress; but the animals who shrink and get shipping fever are the ones who survive. Others die before reaching their destination, or arrive with broken limbs or other injuries. In 1986, USDA inspectors condemned over 7,400 cattle, 3,100 calves, and 5,500 pigs because they were dead or seriously injured before they reached the slaughterhouse, while 570,000 cattle, 57,000 calves, and 643,000 pigs were injured severely enough for parts of their bodies to be condemned.[133]

Animals who die in transit do not die easy deaths. They freeze to death in winter and collapse from thirst and heat exhaustion in summer. They die, lying unattended in stockyards, from injuries sustained in falling off a slippery loading ramp. They suffocate when other animals pile on top of them in overcrowded, badly loaded trucks. They die from thirst or starve when careless stockmen forget to give them food or water. And they die from the sheer stress of the whole terrifying experience. The animal that you may be having for dinner tonight did not die in any of these ways; but these deaths are and always have been part of the overall process that provides people with their meat.

Killing an animal is in itself a troubling act. It has been said that if we had to kill our own meat we would all be vegetarians. Certainly very few people ever visit a slaughterhouse, and films of slaughterhouse operations are not popular on television. People may hope that the meat they buy came from an animal who died without pain, but they do not really want to know about it. Yet those who, by their purchases, require animals to be killed do not deserve to be shielded from this or any other aspect of the production of the meat they buy.

Death, though never pleasant, need not be painful. If all goes according to plan, in developed nations with humane slaughter laws, death comes quickly and painlessly. The animals are supposed to be stunned by electric current or a captive-bolt pistol and have their throats cut while they are still unconscious. They may feel terror shortly before their death, when being goaded up the ramp to slaughter, smelling the blood of those who have gone before; but the moment of death itself can, in theory, be entirely painless. Unfortunately, there is often a gap between theory and practice. A *Washington Post* reporter recently described a slaughterhouse in Virginia operated by Smithfield, the largest meatpacking concern on the United States East Coast:

> The pork process ends in a highly automated state-of-the-art factory where neatly packaged airtight plastic packets of sliced bacon and ham roll off the conveyor belt. But it begins outside behind the plant, in a stinking, muddy, bloodstained pig pen. Inside the Gwaltney of Smithfield slaughterhouse visitors are allowed only a few minutes' stay lest the stench of dead pigs cling to their clothes and bodies long after the visit has ended.
>
> The process begins when the squealing hogs are corralled from their pens up a wooden plank where a worker stuns their heads with an electric shock. As they fall from the shock, a worker quickly hangs the pigs upside-down on a conveyor belt, placing their rear legs in a metal clamp.

> Sometimes the stunned hogs fall off the conveyor belt and regain consciousness, and workers have to scramble to hoist the hogs' legs back into the metal clamps before they begin running wildly through the confined area. The hogs actually are killed by a worker who stabs the stunned and often still-squirming animals with a knife in the jugular vein and lets most of the blood drain out. The freshly butchered pigs then move from the blood-splattered slaughterhouse into the scalding pot.[134]

Much of the suffering that occurs in slaughterhouses is a result of the frantic pace at which the killing line must work. Economic competition means that slaughterhouses strive to kill more animals per hour than their competitors. Between 1981 and 1986, for example, conveyor speed at one large American plant increased from 225 bodies an hour to 275. The pressure to work faster means that less care is taken—and not only with the animals. In 1988 a United States congressional committee reported that no other U.S. industry has a higher injury or illness rate than the slaughter industry. Evidence was given that 58,000 slaughterhouse employees are injured a year, or about 160 a day. If so little care is taken with humans, what is likely to be the fate of the animals? Another major problem with the industry is that, because it is so unpleasant, employees do not last long, and annual turnover rates between 60 and 100 percent are common in many plants. This means a constant stream of inexperienced staff handling frightened animals in a strange environment.[135]

In Britain, where slaughterhouses are in theory tightly controlled by humane slaughter legislation, the government's Farm Animal Welfare Council investigated slaughterhouses and found:

> We have concluded that unconsciousness and insensibility are being assumed to exist in many slaughtering operations when it is highly probable that the degree is not sufficient to render the animal insensitive to pain.

The council added that while there were laws requiring that stunning be conducted effectively and without unnecessary pain by skilled personnel using proper equipment, "we are not satisfied that they are adequately enforced."[136]

Since that report was published, a senior British scientist has raised doubts about whether electrical stunning is painless, even when properly administered. Dr. Harold Hillman, reader in physiology and director of the Unity Laboratory in Applied Neurobiology at the University of Surrey, notes that people who have experienced electric shock, either accidentally or during electroconvulsive therapy for mental illness, report great pain. It is significant, he points out, that electroconvulsive therapy is now normally administered under a general anesthetic. If electric shock instantly rendered the patient incapable of feeling pain, this would not be necessary. For this reason, Dr. Hillman doubts that electrocution, used as a method of capital punishment in some American states, is humane; the prisoner in the electric chair may for a time be paralyzed, but not unconscious. Dr. Hillman then turns to electric stunning in slaughterhouses: "Stunning is believed to be humane, because it is thought that the animals do not suffer pain or distress. This is almost certainly untrue, for the same reasons as have been indicated for the electric chair."[137] So it is quite possible that slaughter is not at all painless, even when properly carried out in a modern slaughterhouse.

Even if these problems could be overcome, there is another problem with the slaughter of animals. Many countries, including Britain and the United States, have an exception for slaughter according to Jewish and Moslem rituals that require the animals to be fully conscious when slaughtered. A second important exception in the United States is that the Federal Humane Slaughter Act, passed in 1958, applies only to slaughterhouses selling meat to the United States government or its agencies and does not apply to the largest number of animals killed—poultry.

Let us consider this second loophole first. There are approximately 6,100 slaughterhouses in the United States, yet fewer than 1,400 were federally inspected for compliance with the humane slaughter law. It is therefore entirely legal for the remaining 4,700 to use the ancient and barbaric poleax; and this method is still in use in some American slaughterhouses.

The poleax is really a heavy sledgehammer rather than an ax. The person wielding the long-handled hammer stands above the animal and tries to knock him unconscious with a single blow. The problem is that the target is moving and the long overhead swing must be carefully aimed; for to succeed the hammer must land at a precise point on the animal's head, and frightened animals are quite likely to move their heads. If the swing is a fraction astray the hammer can crash through the animal's eye or nose; then, as the animal thrashes around in agony and terror, several more blows may be needed to knock the animal unconscious. The most skilled wielder of the poleax cannot be expected to land the blow perfectly every time. As the job may require the killing of eighty or more animals an hour, if the poleax misses in only one out of every hundred swings, the result will still be terrible pain for several animals every day. It should also be remembered that to develop skill with the poleax it is necessary for an unskilled person to get a lot of practice. The practice will be on live animals.

Why are such primitive methods, universally condemned as inhumane, still in use? The reason is the same as in other aspects of animal raising: if humane procedures cost more or reduce the number of animals that can be killed per hour, a firm cannot afford to adopt humane methods while its rivals continue to use the old methods. The cost of the charge used to fire the captive-bolt pistol, though only a few cents per animal, is sufficient to deter slaughterhouses from using it. Electrical stunning is cheaper in the long run, but installation is expensive. Unless the law forces slaughterers to adopt one of these methods they may not be used.

The other major loophole in the humane slaughter laws is that slaughter according to a religious ritual need not comply with the provision that the animal be stunned before being killed. Orthodox Jewish and Moslem dietary laws forbid the consumption of meat from an animal who is not "healthy and moving" when killed. Stunning, which is thought to cause injury prior to cutting the throat, is therefore unacceptable. The idea behind these requirements may have been to prohibit the eating of flesh from an animal who had been found sick or dead; as interpreted by the religiously orthodox today, however, the law also rules out making the animal unconscious a few seconds before it is killed. The killing is supposed to be carried out with a single cut with a sharp knife, aimed at the jugular veins and the carotid arteries. At the time this method of slaughter was laid down in Jewish law it was probably more humane than any alternative; now, however, it is less humane, under the best circumstances, than, for example, the use of the captive-bolt pistol to render an animal instantly insensible.

Moreover, in the United States there are special circumstances that turn this method of slaughter into a grotesque travesty of any humane intentions that may once have lain behind it. This is the result of a combination of the requirements of ritual slaughter and of the Pure Food and Drug Act of 1906, which for sanitary reasons stipulates that a slaughtered animal must not fall in the blood of a previously slaughtered animal. Effectively, this means that the animal must be killed while being suspended from a conveyor belt, or held above the floor in some other way, instead of when lying on the slaughterhouse floor. The requirement does not affect the welfare of an animal who has been made fully unconscious before being killed, since the suspension does not take place until the animal is unconscious; but it has horrible results if the animal must be conscious when killed. Instead of being quickly knocked to the floor and killed almost as they hit the ground, animals being ritually slaughtered in the United States may be shackled around a rear leg, hoisted into the air, and then hang, fully conscious, upside down on the conveyor belt for between two and five minutes—and occasionally much longer if something goes wrong on the "killing line"—before the slaughterer makes his cut. The process has been described as follows:

> When a heavy iron chain is clamped around the leg of a heavy beef animal weighing between 1,000 and 2,000 pounds, and the steer is jerked off its feet, the skin will open and slip away from the bone. The canon bone will often be snapped or fractured.[138]

The animal, upside down, with ruptured joints and often a broken leg, twists frantically in pain and terror, so that it must be gripped by the neck or have a clamp inserted in its nostrils to enable the slaughterer to kill the animal with a single stroke, as the religious law prescribes. It is difficult to imagine a clearer example of how sticking strictly to the letter of a law can pervert its spirit. (It should be noted, however, that even Orthodox rabbis are not unanimous in supporting the prohibition of stunning prior to killing; in Sweden, Norway, and Switzerland, for example, the rabbis have accepted legislation requiring the stunning with no exemptions for ritual slaughter. Many Moslems have also accepted stunning prior to slaughter.[139])

The American Society for the Prevention of Cruelty to Animals has developed a "casting pen" that permits a conscious animal to be killed in compliance with U.S. hygiene regulations without being hoisted by the leg. This device is now used for approximately 80 percent of large cattle undergoing ritual slaughter, but for less than 10 percent of calves. Temple Grandin of Grandin Livestock Handling Systems, Inc., says: "Since religious slaughter is exempt from the Humane Slaughter Act, some plants are not willing to spend money for humaneness."[140]

Those who do not follow Jewish or Moslem dietary laws may believe that the meat they buy has not been killed in this obsolete fashion; but they could be mistaken. For meat to be passed as "kosher" by the Orthodox rabbis, it must, in addition to being from an animal killed while conscious, have had the forbidden tissues, such as veins, lymph nodes, and the sciatic nerve and its branches removed. Cutting these parts out of the hindquarters of an animal is a laborious business and so only the forequarters are sold as kosher meat, and the remainder usually ends up on supermarket shelves without any indication of its origin. This means that far more animals are slaughtered without prior stunning than would be necessary to supply the demand for this type of meat. Britain's Farm Animal Welfare Council has estimated that "a high proportion" of the meat slaughtered by ritual methods is distributed to the open markets.[141]

The slogan "religious freedom" and the charge that those who attack ritual slaughter are motivated by anti-Semitism have sufficed to prevent legislative interference with this practice in the United States, Britain, and many other countries. But obviously one does not have to be anti-Semitic or anti-Moslem to oppose what is done to animals in the name of religion. It is time for adherents of both these religions to consider again whether the current interpretations of laws relating to slaughter are really in keeping with the spirit of religious teaching on compassion. Meanwhile, those who do not wish to eat meat slaughtered contrary to the current teachings of their religion have a simple alternative: not to eat meat at all. In making this suggestion, I am not asking more of religious believers than I ask of myself; it is only that the reasons for them to do it are stronger because of the additional suffering involved in producing the meat they eat.

We live in a time of conflicting currents. While there are those who insist on continuing to kill animals by biblical methods of slaughter, our scientists are busy developing revolutionary techniques by which they hope to change the very nature of the animals themselves. A momentous step toward a world of animals designed by human beings was taken in 1988 when the United States Patent and Trademark Office granted researchers at Harvard University a patent for a genetically engineered mouse, specially made to be more susceptible to cancer so that it can be used to screen possible carcinogens. The grant followed a 1980 Supreme Court decision which made it possible to patent man-made microorganisms, but this was the first time that a patent had been given for an animal.[142]

Religious leaders, animal rights advocates, environmentalists, and ranchers (who are worried about the prospect of being forced to pay royalties to remain competitive) have now formed a coalition to stop the patenting of animals. Meanwhile genetic engineering companies are already working with agribusiness interests to invest money in research designed to create new animals. Unless public pressure puts a stop to such work, there will be huge fortunes to be made from animals who put on more weight or produce more milk or eggs in a shorter time.

The threat to animal welfare is already obvious. Researchers at the U.S. Department of Agriculture's farm at Beltsville, Maryland, have introduced genes for growth hormones into pigs. The genetically altered pigs developed serious side effects, including pneumonia, internal bleeding, and a severe form of crippling arthritis. Apparently only one of these pigs survived to adulthood, and then lived for only two years. This pig was shown on British television, appropriately enough, on *The Money Programme*. The pig was unable to stand.[143] One of the researchers responsible told *The Washington Times*:

> We're at the Wright Brothers stage compared to the 747. We're going to crash and burn for a number of years and not get very far off the ground for a while.

But it will be the animals who "crash and burn," not the researchers. *The Washington Times* also quoted defenders of genetic engineering as rejecting the animal welfare arguments, saying:

> People have cross-bred, domesticated, slaughtered and otherwise exploited animals for centuries. Nothing will change fundamentally.[144]

As this chapter has shown, that is true. We have long treated animals as things for our convenience, and for the last thirty years we have been applying our latest scientific techniques to make them serve our ends better. Genetic engineering, revolutionary as it may be in one sense, is in another sense just one more way of bending animals to our purposes. The real need is that attitudes and practices change fundamentally.

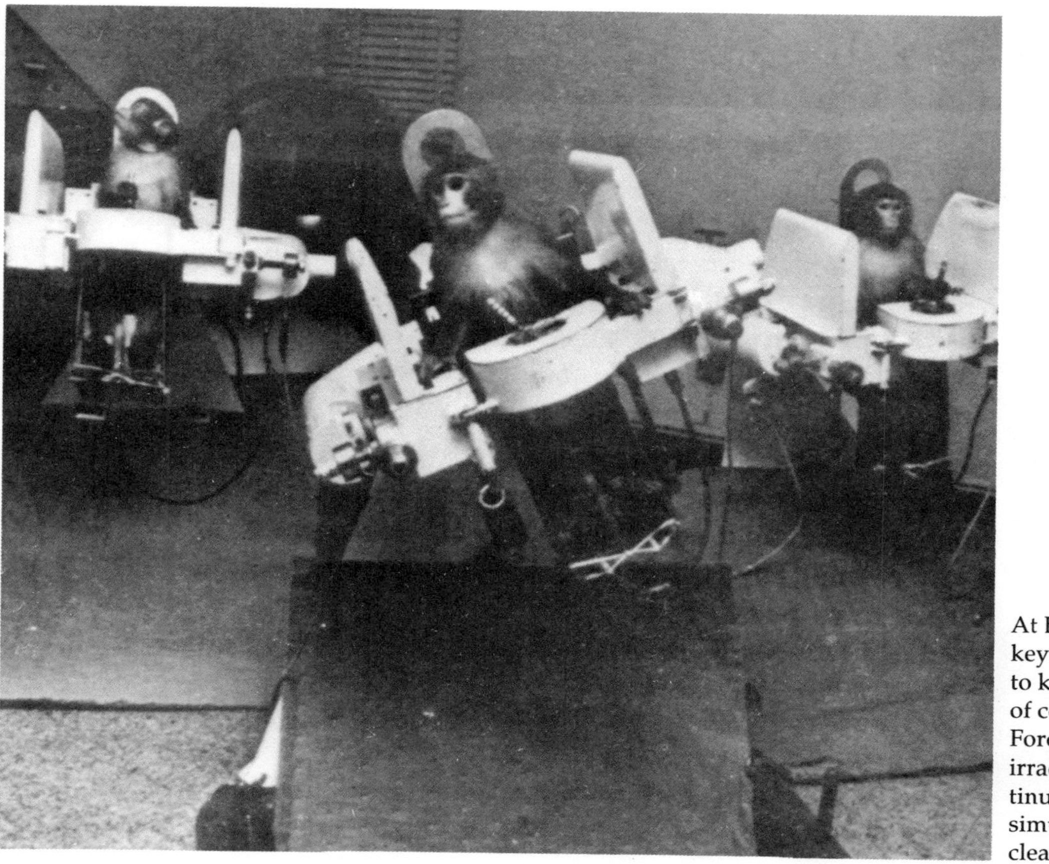

At Brooks Air Force Base in Texas, monkeys are trained through electric shocks to keep these platforms level by means of controls that simulate the flying of air Force bombers. They are then gassed or irradiated to test how long they can continue to keep the platforms level under simulated conditions of chemical or nuclear attack. See pp. 25–28.

A rhesus monkey confined to a treadmill at the Armed Forces Radiobiology Research Institute, Bethesda, Maryland, which conducts military research on lethal doses of gamma-neutron radiation. The primates are trained through electric shock to turn a wheel at a speed between one and five miles per hour. Following an eight-week conditioning period, the monkeys are irradiated and again tested in the wheel until death. The research compares the individual monkey's performance pattern before and after exposure to lethal radiation. See pp. 29–31. (Photo by Henry Spira.)

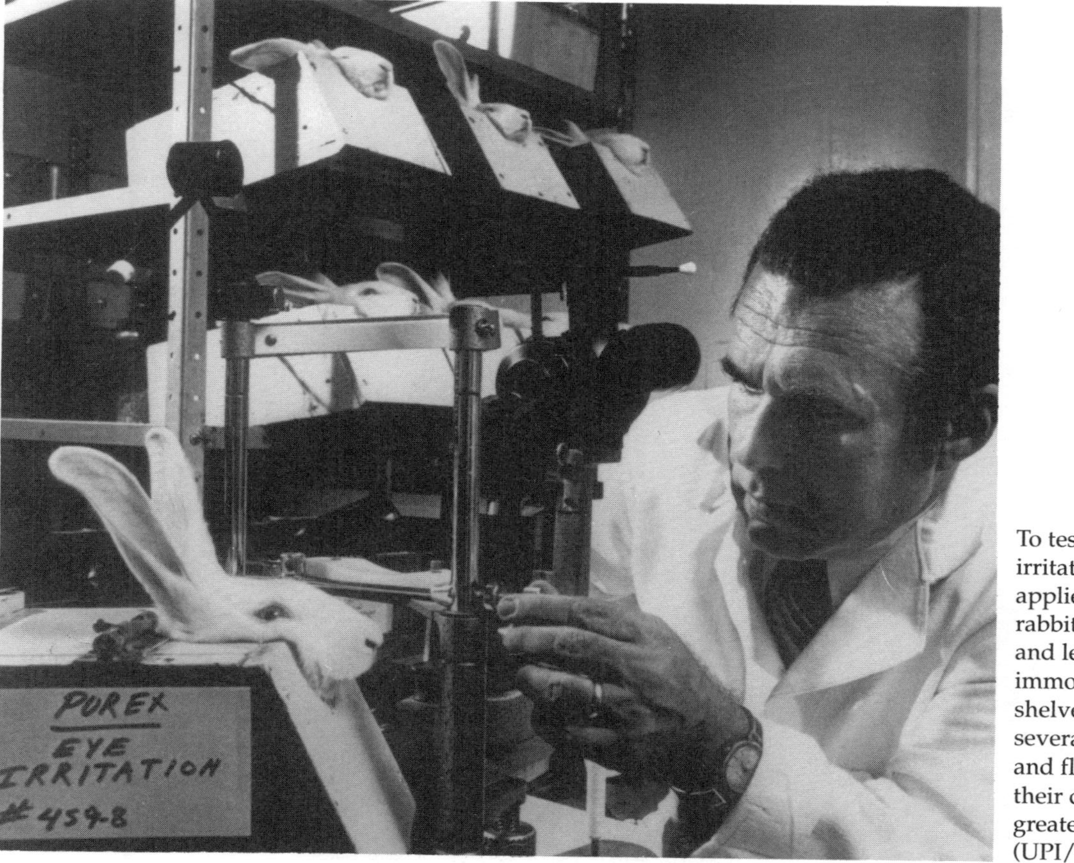

To test a detergent for possible eye irritation, a paste of the detergent is applied directly to the eyes of rabbits, which are then bandaged and left to react to the paste. The immobilized rabbits are stacked on shelves (visible in the background) for several hours. Since rabbits cannot cry and flush the detergent from the eye, their capacity for irritation is much greater than that of humans. (UPI/Bettmann Archives Photo.)

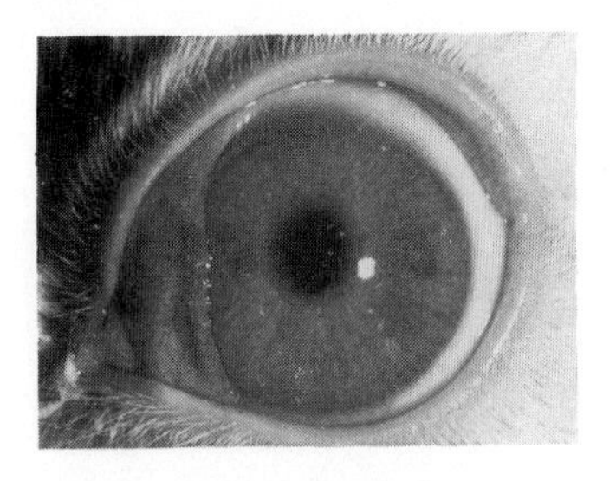
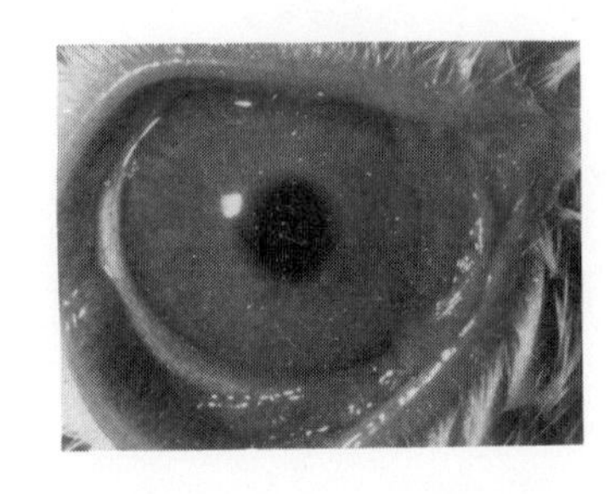
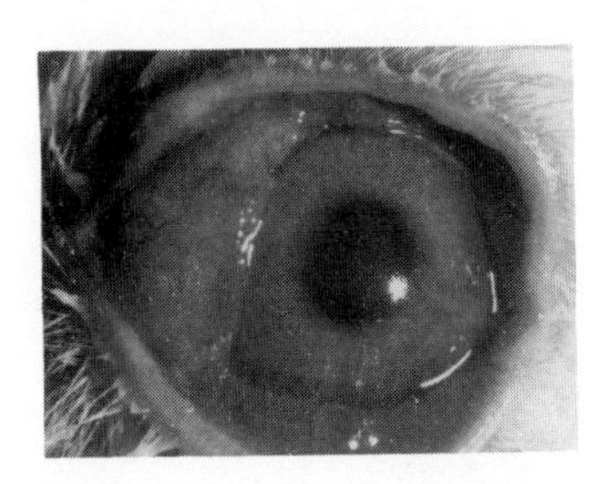
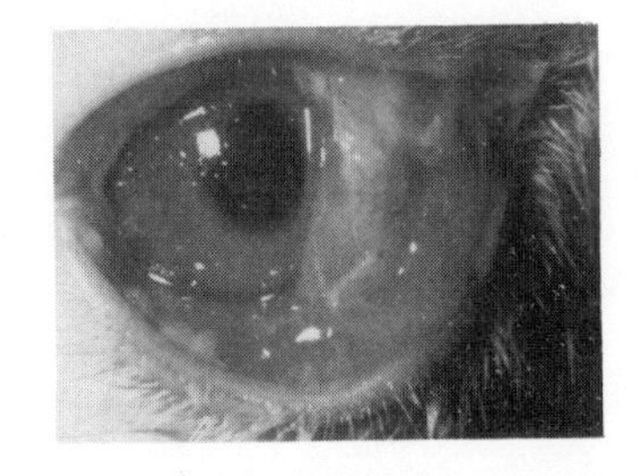
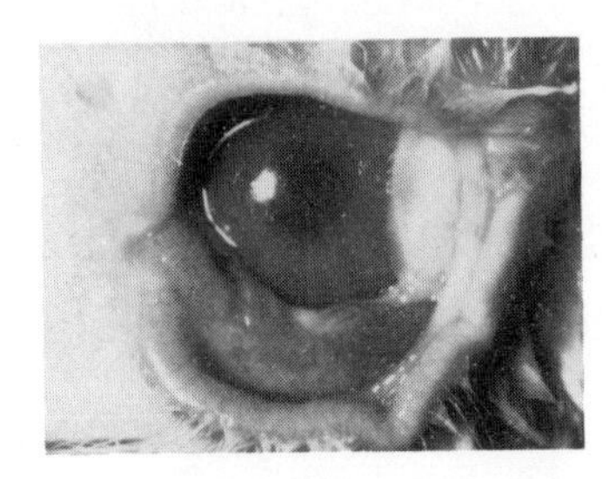
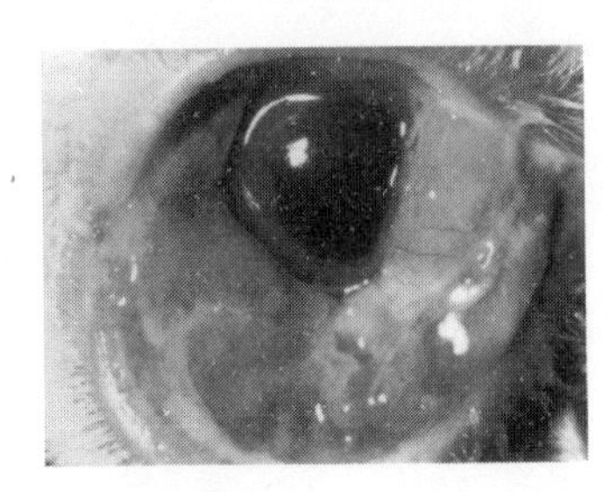

The above photographs show the effects of irritants placed in the eyes of rabbits as part of the Draize Test. They are taken from the U.S. Consumer Product Safety Commission's *Illustrated Guide for Grading Eye Irritation Caused by Hazardous Substances.* According to the introduction to this publication, its stated purpose is "to assist in training laboratory personnel ... and thereby contribute to more uniform interpretations of the results obtained when a substance is tested in accordance with the official method." In other words, laboratory staff are expected to place potentially irritating substances in the eyes of rabbits, wait for periods ranging from a few hours to seven days, and then judge the irritancy of these substances by comparing the appearance of the rabbits' eyes with these photographs.

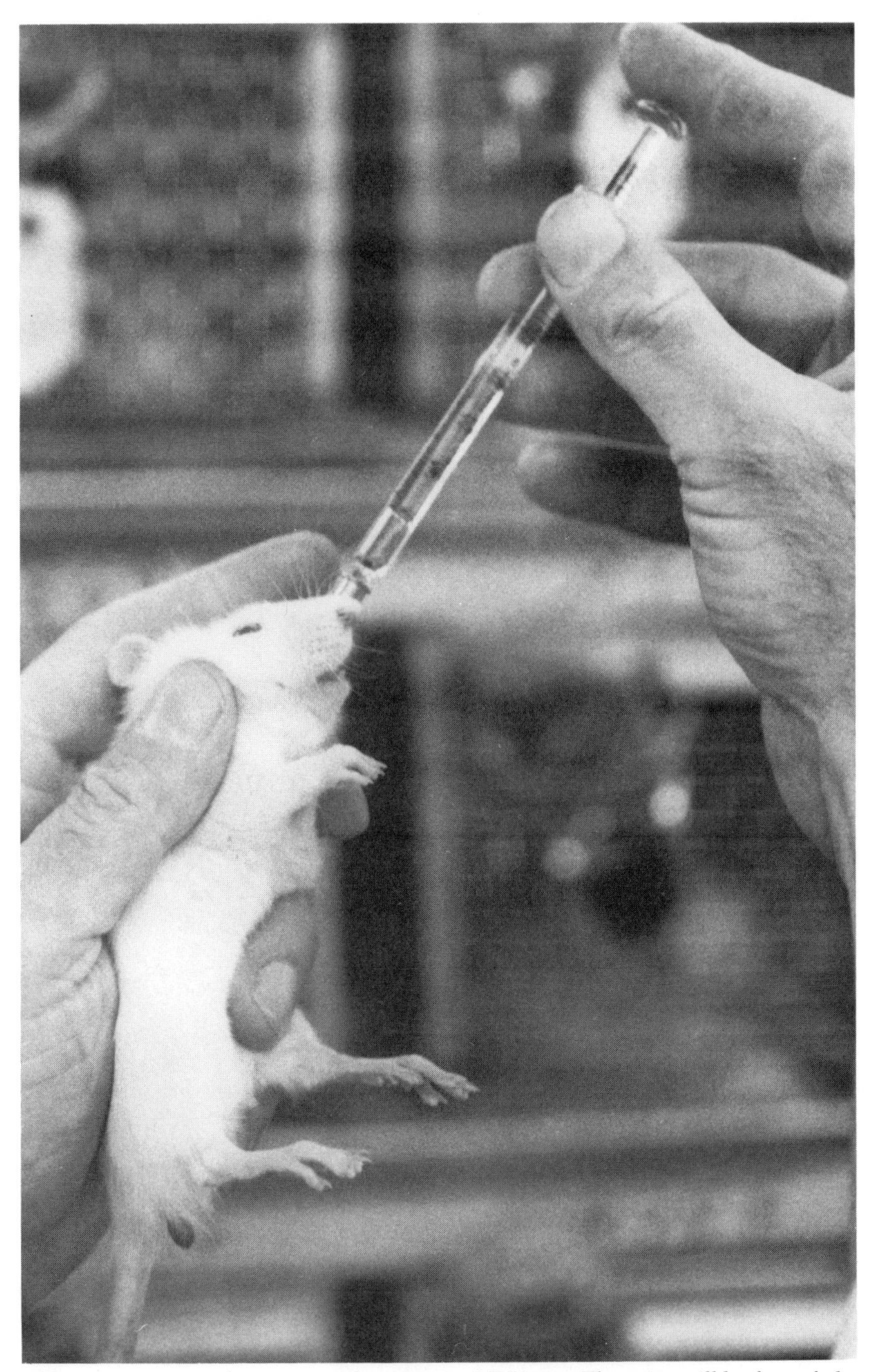

This mouse is one of a group undergoing an LD50 test. The mice will be force-fed the substance to be tested (perhaps a food coloring or synthetic flavoring agent) until 50 percent of the group are poisoned to death. See pp. 53–56.

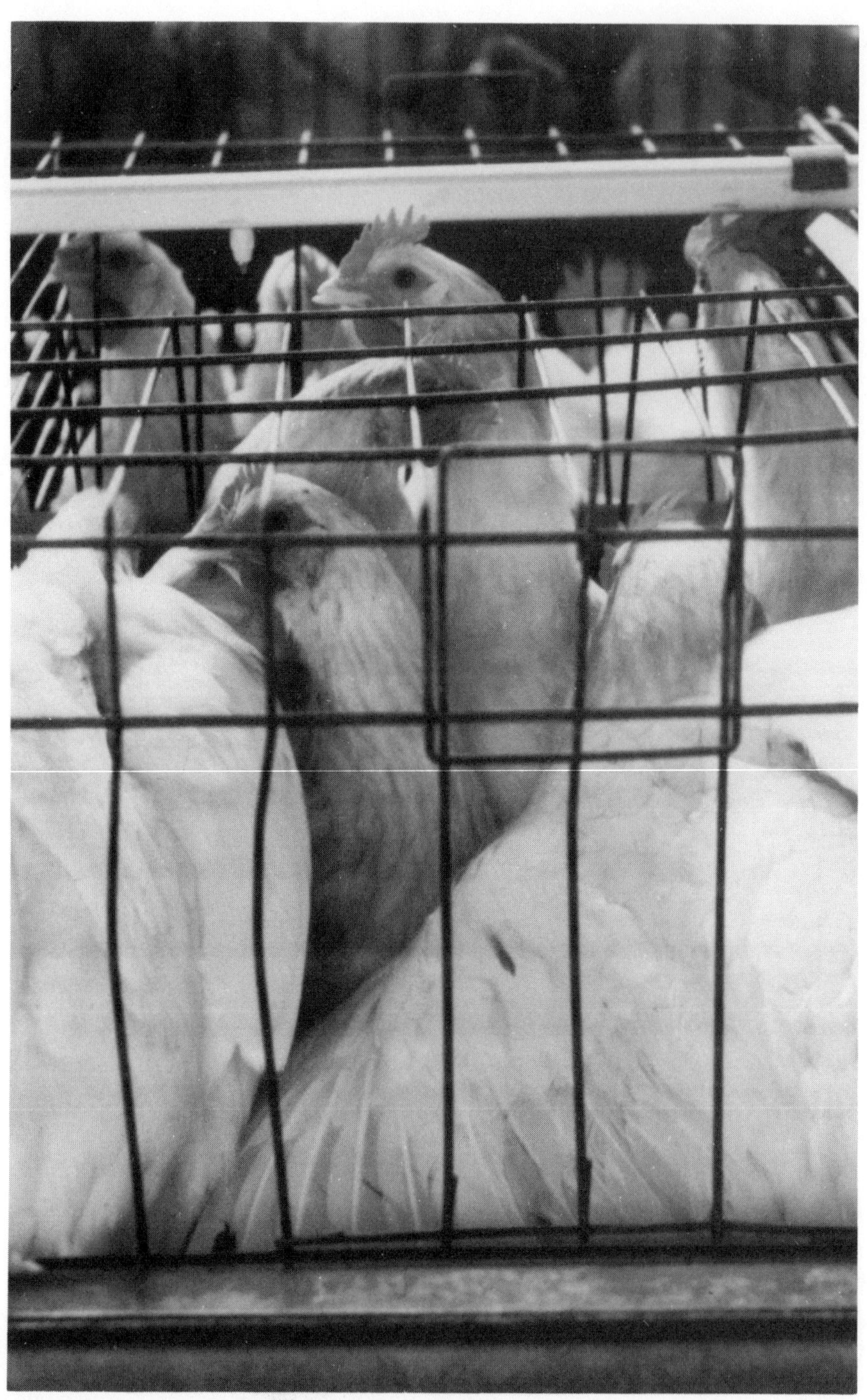

Closeup of battery cage at Somerset Poultry Farm, Victoria, Australia. There were seven hens in this cage, which measured about 18 inches by 18 inches. (Photo by Patty Mark.)

A production line carries live chickens on their way to slaughter in the killing room of a processing plant. (Photo by Jim Mason and J.A. Keller, from the book *Animal Factories*.)

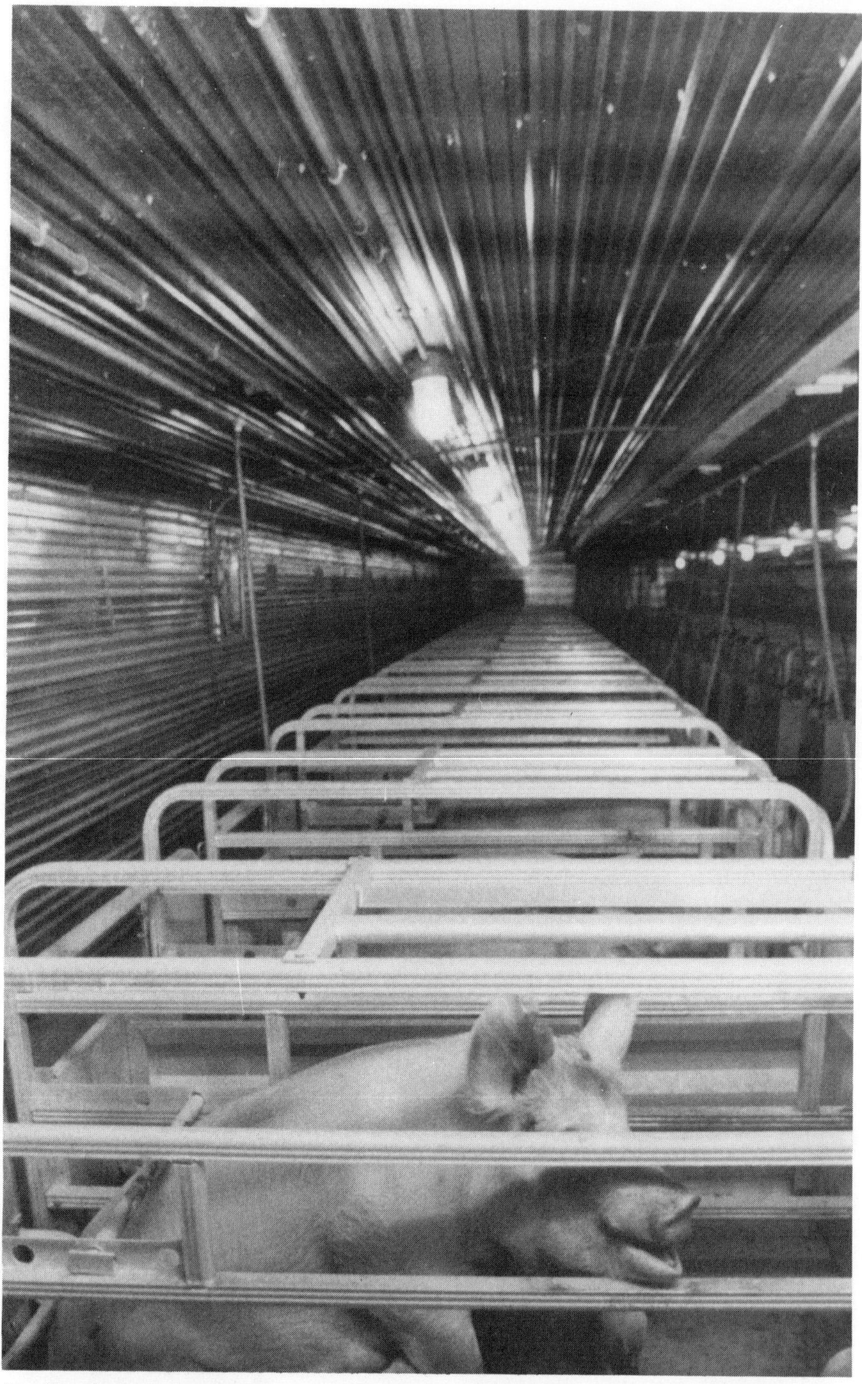

During their pregnancies, sows are confined in stalls that do not permit them to turn around or walk to and fro. (Photo by Jim Mason and J.A. Keller, from the book *Animal Factories*.)

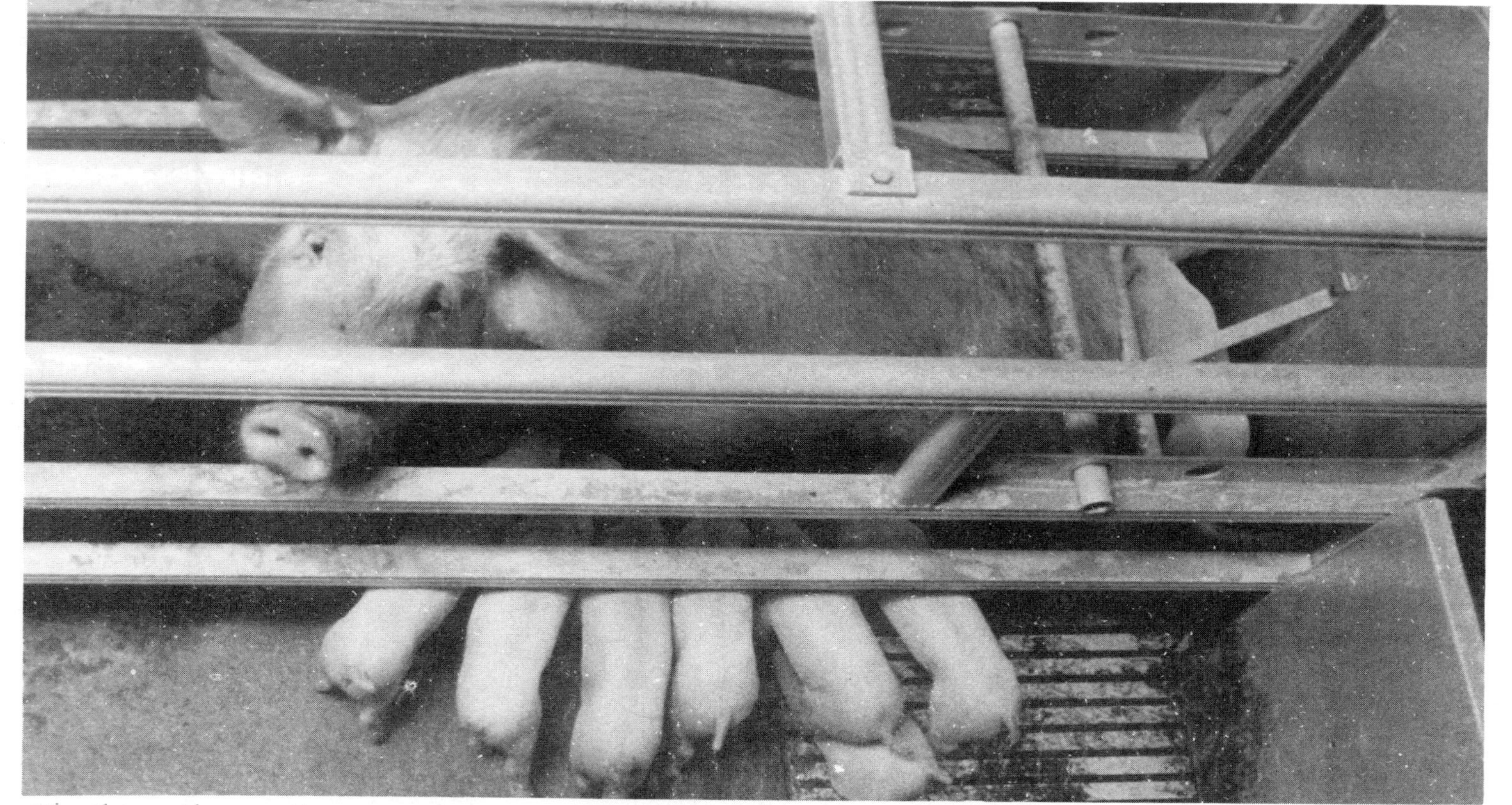

After their confinement during pregnancy, sows are often immobilized from the time they give birth until the piglets are weaned. (Photo by Jim Mason and J.A. Keller, from the book *Animal Factories*.)

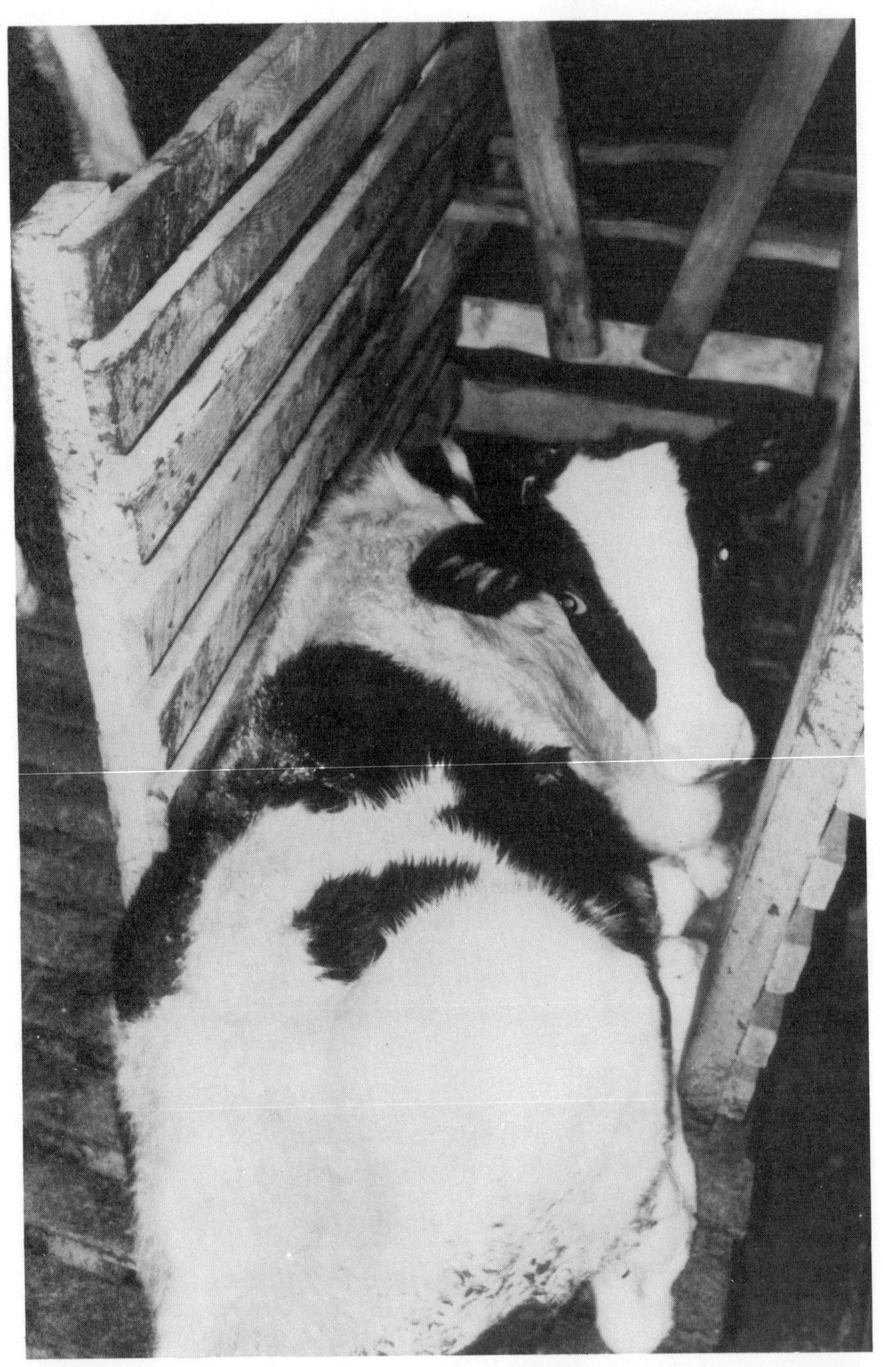

This veal calf will spend its whole life constrained to this pen so that its anemic muscles will remain tender when butchered. In order to lie down, such calves must hunch up to fit their legs into the 22-inch stall. (Photo courtesy of Humane Farming Association.)

Chapter 4

Becoming a Vegetarian . . .

or how to produce less suffering and more food at a reduced cost to the environment

Now that we have understood the nature of speciesism and seen the consequences it has for nonhuman animals it is time to ask: What can we do about it? There are many things that we can and should do about speciesism. We should, for instance, write to our political representatives about the issues discussed in this book; we should make our friends aware of these issues; we should educate our children to be concerned about the welfare of all sentient beings; and we should protest publicly on behalf of nonhuman animals whenever we have an effective opportunity to do so.

While we should do all these things, there is one other thing we can do that is of supreme importance; it underpins, makes consistent, and gives meaning to all our other activities on behalf of animals. This one thing is that we take responsibility for our own lives, and make them as free of cruelty as we can. The first step is that we cease to eat animals. Many people who are opposed to cruelty to animals draw the line at becoming a vegetarian. It was of such people that Oliver Goldsmith, the eighteenth-century humanitarian essayist, wrote: "They pity, and they eat the objects of their compassion."[1]

As a matter of strict logic, perhaps, there is no contradiction in taking an interest in animals on both compassionate and gastronomic grounds. If one is opposed to inflicting suffering on animals, but not to the painless killing of animals, one could consistently eat animals who had lived free of all suffering and been instantly, painlessly slaughtered. Yet practically and psychologically it is impossible to be consistent in one's concern for nonhuman animals while continuing to dine on them. If we are prepared to take the life of another being merely in order to satisfy our taste for a particular type of food, then that being is no more than a means to our end. In time we will come to regard pigs, cattle, and chickens as things for us to use, no matter how strong our compassion may be; and when we find that to continue to obtain supplies of the bodies of these animals at a price we are able to pay it is necessary to change their living conditions a little, we will be unlikely to regard these changes too critically. The factory farm is nothing more than the application of technology to the idea that animals are means to our ends. Our eating habits are dear to us and not easily altered. We have a strong interest in convincing ourselves that our concern for other animals does not require us to stop eating them. No one in the habit of eating an animal can be completely without bias in judging whether the conditions in which that animal is reared cause suffering.

It is not practically possible to rear animals for food on a large scale without inflicting considerable suffering. Even if intensive methods are not used, traditional farming involves castration, separation of mother and young, breaking up social groups, branding, transportation to the slaughterhouse, and finally slaughter itself. It is difficult to imagine how animals could be reared for food without these forms of suffering. Possibly it could be done on a small scale, but we could never feed today's huge urban populations with meat raised in this manner. If it could be done at all, the animal flesh thus produced would be vastly more expensive than animal flesh is today—and rearing animals is already an expensive and inefficient way of producing protein. The flesh of animals reared and killed with equal consideration for the welfare of animals while they were alive would be a delicacy available only to the rich.

All this is, in any case, quite irrelevant to the immediate question of the ethics of our daily diet. Whatever the theoretical possibilities of rearing animals without suffering may be, the fact is that the meat available from butchers and supermarkets comes from animals who were not treated with any real consideration at all while being reared. So we must ask ourselves, not: Is it *ever* right to eat meat? but: Is it right to eat *this* meat? Here I think that those who are opposed to the needless killing of animals and those who oppose only the infliction of suffering must join together and give the same, negative answer.

Becoming a vegetarian is not merely a symbolic gesture. Nor is it an attempt to isolate oneself from the ugly realities of the world, to keep oneself pure and so without responsibility for the cruelty and carnage all around. Becoming a vegetarian is a highly practical and effective step one can take toward ending both the killing of nonhuman animals and the infliction of suffering upon them. Assume, for the moment, that it is only the suffering that we disapprove of, not the killing. How can we stop the use of the intensive methods of animal rearing described in the previous chapter?

So long as people are prepared to buy the products of intensive farming, the usual forms of protest and political action will never bring about a major reform. Even in supposedly animal-loving Britain, although the wide controversy stirred by the publication of Ruth Harrison's *Animal Machines* forced the government to appoint a group of impartial experts (the Brambell committee) to investigate the issue of mistreatment of animals and make recommendations, when the committee reported the government refused to carry out its recommendations. In 1981 the House of Commons Agriculture Committee made yet another inquiry into intensive farming, and this inquiry also led to recommendations for eliminating the worst abuses. Once again, nothing was done.[2] If this was the fate of the movement for reform in Britain, nothing better can be expected in the United States, where the agribusiness lobby is still more powerful.

This is not to say that the normal channels of protest and political action are useless and should be abandoned. On the contrary, they are a necessary part of the overall struggle for effective change in the treatment of animals. In Britain, especially, organizations like Compassion in World Farming have kept the issue before the public, and even succeeded in bringing about an end to veal crates. More recently American groups have also started to arouse public concern over intensive farming. But in themselves, these methods are not enough.

The people who profit by exploiting large numbers of animals do not need our approval. They need our money. The purchase of the corpses of the animals they rear is the main support the factory farmers ask from the public (the other, in many countries, is big government subsidies). They will use intensive methods as long as they can sell what they produce by these methods; they will have the resources needed to fight reform politically; and they will be able to defend themselves against criticism with the reply that they are only providing the public with what it wants.

Hence the need for each one of us to stop buying the products of modern animal farming—even if we are not convinced that it would be wrong to eat animals who have lived pleasantly and died painlessly. Vegetarianism is a form of boycott. For most vegetarians the boycott is a permanent one, since once they have broken away from flesh-eating habits they can no longer approve of slaughtering animals in order to satisfy the trivial desires of their palates. But the moral obligation to boycott the meat available in butcher shops and supermarkets today is just as inescapable for those who disapprove only of inflicting suffering, and not of killing. Until we boycott meat, and all other products of animal factories, we are, each one of us, contributing to the continued existence, prosperity, and growth of factory farming and all the other cruel practices used in rearing animals for food.

It is at this point that the consequences of speciesism intrude directly into our lives, and we are forced to attest personally to the sincerity of our concern for nonhuman animals. Here we have an opportunity to do something, instead of merely talking and wishing the politicians would do something. It is easy to take a stand about a remote issue, but speciesists, like racists, reveal their true nature when the issue comes nearer home. To protest about bullfighting in Spain, the eating of dogs in South Korea, or the slaughter of baby seals in Canada while continuing to eat eggs from hens who have spent their lives crammed into cages, or veal from calves who have been deprived of their mothers, their proper diet, and the freedom to lie down with their legs extended, is like denouncing apartheid in South Africa while asking your neighbors not to sell their houses to blacks.

To make the boycott aspect of vegetarianism more effective, we must not be shy about our refusal to eat flesh. Vegetarians in omnivorous societies are always being asked about the reasons for their strange diets. This can be irritating, or even embarrassing, but it also provides opportunities to tell people about cruelties of which they may be unaware. (I first learned of the existence of factory farming from a vegetarian who took the time to explain to me why he wasn't eating the same food I was.) If a boycott is the only way to stop cruelty, then we must encourage as many as possible to join the boycott. We can only be effective in this if we ourselves set the example.

People sometimes attempt to justify eating flesh by saying that the animal was already dead when they bought it. The weakness of this rationalization—which I have heard used, quite seriously, many times—should be obvious as soon as we consider vegetarianism as a form of boycott. The nonunion grapes available in stores during the grape boycott inspired by Cesar Chavez's efforts to improve the wages and conditions of the grape-pickers had already been produced by underpaid laborers, and we could no more raise the pay those laborers had received for picking those grapes than we could bring our steak back to life. In both cases the aim of the boycott is not to alter the past but to prevent the continuation of the conditions to which we object.

I have emphasized the boycott element of vegetarianism so much that the reader may ask whether, if the boycott does not spread and prove effective, anything has been achieved by becoming a vegetarian. But we must often venture when we cannot be certain of success, and it would be no argument against becoming a vegetarian if this were all that could be said against it, since none of the great movements against oppression and injustice would have existed if their leaders had made no efforts until they were assured of success. In the case of vegetarianism, however, I believe we do achieve something by our individual acts, even if the boycott as a whole should not succeed. George Bernard Shaw once said that he would be followed to his grave by numerous sheep, cattle, pigs, chickens, and a whole shoal of fish, all grateful at having been spared from slaughter because of his vegetarian diet. Although we cannot identify any individual animals whom we have benefited by becoming a vegetarian, we can assume that our diet, together with that of the many others who are already avoiding meat, will have some impact on the number of animals raised in factory farms and slaughtered for food. This assumption is reasonable because the number of animals raised and slaughtered depends on the profitability of this process, and this profit depends in part on the demand for the product. The smaller the demand, the lower the price and the lower the profit. The lower the profit, the fewer the animals that will be raised and slaughtered. This is elementary economics, and it can easily be observed in tables published by the poultry trade journals, for instance, that there is a direct correlation between the price of poultry and the number of chickens placed in broiler sheds to begin their joyless existence.

So vegetarianism is really on even stronger ground than most other boycotts or protests. The person who boycotts South African produce in order to bring down apartheid achieves nothing unless the boycott succeeds in forcing white South Africans to modify their policies (though the effort may have been well worth making, whatever the outcome); but vegetarians know that they do, by their actions, contribute to a reduction in the suffering and slaughter of animals, whether or not they live to see their efforts spark off a mass boycott of meat and an end to cruelty in farming.

In addition to all this, becoming a vegetarian has a special significance because the vegetarian is a practical, living refutation of a common, yet utterly false, defense of factory farming methods. It is sometimes said that these methods are needed to feed the world's soaring population. Because the truth here is so important—important enough, in fact, to amount to a convincing case for vegetarianism that is quite independent of the question of animal welfare that I have emphasized in this book—I shall digress briefly to discuss the fundamentals of food production.

At this moment, millions of people in many parts of the world do not get enough to eat. Millions more get a sufficient quantity, but they do not get the right kind of food; mostly, they do not get enough protein. The question is, does raising food by the methods practiced in the affluent nations make a contribution to the solution of the hunger problem?

Every animal has to eat in order to grow to the size and weight at which it is considered ready for human beings to eat. If a calf, say, grazes on rough pasture land that grows only grass and could not be planted with corn or any other crop that provides food edible by human beings, the result will be a net gain of protein for human beings, since the grown calf provides us with protein that we cannot—yet—extract economically from grass. But if we take that same calf and place him in a feedlot, or any other confinement system, the picture changes. The calf must now be fed. No matter how little space he and his companions are crowded into, land must be used to grow the corn, sorghum, soybeans, or whatever it is that the calf eats. Now we are feeding the calf food that we ourselves could eat. The calf needs most of this food for the ordinary physiological processes of day-to-day living. No matter how severely the calf is prevented from exercising, his body must still burn food merely to keep him alive. The food is also used to build inedible parts of the calf's body, like bones. Only the food left over after these needs are satisfied can be turned into flesh, and eventually be eaten by human beings.

How much of the protein in his food does the calf use up, and how much is available for human beings? The answer is surprising. It takes twenty-one pounds of protein fed to a calf to produce a single pound of animal protein for humans. We get back less than 5 percent of what we put in. No wonder that Frances Moore Lappé has called this kind of farming "a protein factory in reverse"![3]

We can put the matter another way. Assume we have one acre of fertile land. We can use this acre to grow a high-protein plant food, like peas or beans. If we do this, we will get between three hundred and five hundred pounds of protein from our acre. Alternatively we can use our acre to grow a crop that we feed to animals, and then kill and eat the animals. Then we will end up with between forty and fifty-five pounds of protein from our acre. Interestingly enough, although most animals convert plant protein into animal protein more efficiently than cattle do—a pig, for instance, needs "only" eight pounds of protein to produce one pound for humans—this advantage is almost eliminated when we consider how much protein we can produce per acre, because cattle can make use of sources of protein that are indigestible for pigs. So most estimates conclude that plant foods yield about ten times as much protein per acre as meat does, although estimates vary, and the ratio sometimes goes as high as twenty to one.[4]

If instead of killing the animals and eating their flesh we use them to provide us with milk or eggs we improve our return considerably. Nevertheless the animals must still use protein for their own purposes and the most efficient forms of egg and milk production do not yield more than a quarter of the protein per acre that can be provided by plant foods.

Protein is, of course, only one necessary nutrient. If we compare the total number of calories produced by plant foods with animal foods, the comparison is still all in favor of plants. A comparison of yields from an acre sown with oats or broccoli with yields from an acre used for feed to produce pork, milk, poultry, or beef shows that the acre of oats produces six times the calories yielded by pork, the most efficient of the animal products. The acre of broccoli yields nearly three times as many calories as pork. Oats produce more than twenty-five times as many calories per acre as beef. Looking at some other nutrients shatters other myths fostered by meat and dairy industries. For instance, an acre of broccoli produces twenty-four times the iron produced by an acre used for beef, and an acre of oats sixteen times the same amount of iron. Although milk production does yield more calcium per acre than oats, broccoli does better still, providing five times as much calcium as milk.[5]

The implications of all this for the world food situation are staggering. In 1974 Lester Brown of the Overseas Development Council estimated that if Americans were to reduce their meat consumption by only 10 percent for one year, it would free at least 12 million tons of grain for human consumption—or enough to feed 60 million people. Don Paarlberg, a former U.S. assistant secretary of agriculture, has said that merely reducing the U.S. livestock population by half would make available enough food to make up the calorie deficit of the nonsocialist underdeveloped nations nearly four times over.[6] Indeed, the food wasted by animal production in the affluent nations would be sufficient, if properly distributed, to end both hunger and malnutrition throughout the world. The simple answer to our question, then, is that raising animals for food by the methods used in the industrial nations does not contribute to the solution of the hunger problem.

Meat production also puts a strain on other resources. Alan Durning, a researcher at the Worldwatch Institute, an environmental thinktank based in Washington, D.C., has calculated that one pound of steak from steers raised in a feedlot costs five pounds of grain, 2,500 gallons of water, the energy equivalent of a gallon of gasoline, and about thirty-five pounds of eroded topsoil. More than a third of North America is taken up with grazing, more than half of U.S. croplands are planted with livestock feed, and more than half of all water consumed in the United States goes to livestock.[7] In all these respects plant foods are far less demanding of our resources and our environment.

Let us consider energy usage first. One might think that agriculture is a way of using the fertility of the soil and the energy provided by sunlight to increase the amount of energy available to us. Traditional agriculture does precisely that. Corn grown in Mexico, for instance, produces 83 calories of food for each calorie of fossil fuel energy input. Agriculture in developed countries, however, relies on a large input of fossil fuel. The most energy-efficient form of food production in the United States (oats, again) produces barely 2.5 food calories per calorie of fossil fuel energy, while potatoes yield just over 2, and wheat and soybeans around 1.5. Even these meager results, however, are a bonanza compared to United States animal production, every form of which costs more energy than it yields. The least inefficient—range-land beef—uses more than 3 calories of fossil fuel for every food calorie it yields; while the most inefficient—feedlot beef—takes 33 fuel calories for every food calorie. In energy efficiency, eggs, lamb, milk, and poultry come between the two forms of beef production. In other words, limiting ourselves to United States agriculture, growing crops is generally at least five times more energy-efficient than grazing cattle, about twenty times more energy-efficient than producing chickens, and more than fifty times as energy-efficient as feedlot cattle production.[8] United States animal production is workable only because it draws on millions of years of accumulated solar energy, stored in the ground as oil and coal. This makes economic sense to agribusiness corporations because meat is worth more than oil; but for a rational long-term use of our finite resources, it makes no sense at all.

Animal production also compares poorly with crop production as far as water use is concerned. A pound of meat requires fifty times as much water as an equivalent quantity of wheat.[9] *Newsweek* graphically described this volume of water when it said, "The water that goes into a 1000 pound steer would float a destroyer."[10] The demands of animal production are drying up the vast underground pools of water on which so many of the drier regions of America, Australia, and other countries rely. In the cattle country that stretches from western Texas to Nebraska, for example, water tables are falling and wells are going dry as the huge underground lake known as the Ogalalla Aquifer—another resource which, like oil and coal, took millions of years to create—continues to be used up to produce meat.[11]

Nor should we neglect what animal production does to the water that it does not use. Statistics from the British Water Authorities Association show that there were more than 3,500 incidents of water pollution from farms in 1985. Here is just one example from that year: a tank at a pig unit burst, sending a quarter-million liters of pig excrement into the River Perry and killing 110,000 fish. More than half of the prosecutions by water authorities for serious pollution of rivers are now against farmers.[12] This is not surprising, for a modest 60,000-bird egg factory produces eighty-two tons of manure every week, and in the same period two thousand pigs will excrete twenty-seven tons of manure and thirty-two tons of urine. Dutch farms produce 94 million tons of manure a year, but only 50 million can safely be absorbed by the land. The excess, it has been calculated, would fill a freight train stretching 16,000 kilometers from Amsterdam to the farthest shores of Canada. But the excess is not being carted away; it is dumped on the land where it pollutes water supplies and kills the remaining natural vegetation in the farming regions of the Netherlands.[13] In the United States, farm animals produce 2 billion tons of manure a year—about ten times that of the human population—and half of it comes from factory-reared animals, where the waste does not return naturally to the land.[14] As one pig farmer put it: "Until fertilizer gets more expensive than labor, the waste has very little value to me."[15] So the manure that should restore the fertility of our soils ends up polluting our streams and rivers.

It will, however, be the squandering of the forests that turns out to be the greatest of all the follies caused by the demand for meat. Historically, the desire to graze animals has been the dominant motive for clearing forests. It still is today. In Costa Rica, Colombia, and Brazil, in Malaysia, Thailand, and Indonesia, rainforests are being cleared to provide grazing land for cattle. But the meat produced from the cattle does not benefit the poor of those countries. Instead it is sold to the well-to-do in the big cities, or it is exported. Over the past twenty-five years, nearly half of Central America's tropical rainforests have been destroyed, largely to provide beef to North America.[16] Perhaps 90 percent of the plant and animal species on this planet live in the tropics, many of them still unrecorded by science.[17] If the clearing continues at its present rate, they will be pushed into extinction. In addition, clearing the land causes erosion, the increased runoff leads to flooding, peasants no longer have wood for fuel, and rainfall may be reduced.[18]

We are losing these forests just at the moment when we are starting to learn how truly vital they are. Since the North American drought of 1988, many people have heard of the threat posed to our planet by the greenhouse effect, caused mainly by increasing amounts of carbon dioxide in the atmosphere. Forests store immense amounts of carbon; it has been estimated that despite all the clearing that has taken place, the world's remaining forests still hold four hundred times the amount of carbon released into the atmosphere each year by human use of fossil fuels. Destroying a forest releases the carbon into the atmosphere in the form of carbon dioxide. Conversely, a new, growing forest absorbs carbon dioxide from the atmosphere and locks it up as living matter. The destruction of existing forests will intensify the greenhouse effect; in large-scale reforestation, combined with other measures to reduce the output of carbon dioxide, lies our only hope of mitigating it.[19] If we fail to do so, the warming of our planet will mean, within the next fifty years, widespread droughts, further destruction of forests from climatic change, the extinction of innumerable species unable to cope with the changes in their habitat, and a melting of the polar ice caps, which will in turn raise sea levels and flood coastal cities and plains. A rise of one meter in the level of the sea would flood 15 percent of Bangladesh, affecting 10 million people; and it would threaten the very existence of some low-lying Pacific island nations such as the Maldives, Tuvalu, and Kiribati.[20]

Forests and meat animals compete for the same land. The prodigious appetite of the affluent nations for meat means that agribusiness can pay more than those who want to preserve or restore the forests. We are, quite literally, gambling with the future of our planet—for the sake of hamburgers.

How far should we go? The case for a radical break in our eating habits is clear; but should we eat nothing but plant foods? Where exactly do we draw the line?

Drawing precise lines is always difficult. I shall make some suggestions, but the reader might well find what I say here less convincing than what I have said before about the more clear-cut cases. You must decide for yourself where you are going to draw the line, and your decision may not coincide exactly with mine. This does not matter all that much. We can distinguish bald men from men who are not bald without deciding every borderline case. It is agreement on the fundamentals that is important.

I hope that anyone who has read this far will recognize the moral necessity of refusing to buy or eat the flesh or other products of animals who have been reared in modern factory farm conditions. This is the clearest case of all, the absolute minimum that anyone with the capacity to look beyond considerations of narrow self-interest should be able to accept.

Let us see what this minimum involves. It means that, unless we can be sure of the origin of the particular item we are buying, we must avoid chicken, turkey, rabbit, pork, veal, beef, and eggs. At the present time relatively little lamb is intensively produced; but some is, and more may be in future. The likelihood of your beef coming from a feedlot or some other form of confinement—or from grazing land created by clearing rainforest—will depend on the country in which you live. It is possible to obtain supplies of all these meats that do not come from factory farms, but unless you live in a rural area this takes a lot of effort. Most butchers have no idea how the animals whose bodies they are selling were raised. In some cases, such as that of chickens, traditional methods of rearing have disappeared so completely that it is almost impossible to buy a chicken that was free to roam outdoors; and veal is a meat that simply cannot be produced humanely. Even when meat is described as "organic" this may mean no more than that the animals were not fed the usual doses of antibiotics, hormones, or other drugs; small solace for an animal who was not free to walk around outdoors. As for eggs, in many countries "free range eggs" are widely available, though in most parts of the United States they are still very difficult to get.

Once you have stopped eating poultry, pork, veal, beef, and factory farm eggs the next step is to refuse to eat any slaughtered bird or mammal. This is only a very small additional step, since so few of the birds or mammals commonly eaten are not intensively reared. People who have no experience of how satisfying an imaginative vegetarian diet can be may think of it as a major sacrifice. To this I can only say: "Try it!" Buy a good vegetarian cookbook (some are listed in Appendix 2 of this book) and you will find that being a vegetarian is no sacrifice at all. The reason for taking this extra step may be the belief that it is wrong to kill these creatures for the trivial purpose of pleasing our palates; or it may be the knowledge that even when these animals are not intensively raised they suffer in the various other ways described in the previous chapter.

Now more difficult questions arise. How far down the evolutionary scale shall we go? Shall we eat fish? What about shrimps? Oysters? To answer these questions we must bear in mind the central principle on which our concern for other beings is based. As I said in the first chapter, the only legitimate boundary to our concern for the interests of other beings is the point at which it is no longer accurate to say that the other being has interests. To have interests, in a strict, nonmetaphorical sense, a being must be capable of suffering or experiencing pleasure. If a being suffers, there can be no moral justification for disregarding that suffering, or for refusing to count it equally with the like suffering of any other being. But the converse of this is also true. If a being is not capable of suffering, or of enjoyment, there is nothing to take into account.

So the problem of drawing the line is the problem of deciding when we are justified in assuming that a being is incapable of suffering. In my earlier discussion of the evidence that nonhuman animals are capable of suffering, I suggested two indicators of this capacity: the behavior of the being, whether it writhes, utters cries, attempts to escape from the source of pain, and so on; and the similarity of the nervous system of the being to our own. As we proceed down the evolutionary scale we find that on both these grounds the strength of the evidence for a capacity to feel pain diminishes. With birds and mammals the evidence is overwhelming. Reptiles and fish have nervous systems that differ from those of mammals in some important respects but share the basic structure of centrally organized nerve pathways. Fish and reptiles show most of the pain behavior that mammals do. In most species there is even vocalization, although it is not audible to our ears. Fish, for instance, make vibratory sounds, and different "calls" have been distinguished by researchers, including sounds indicating "alarm" and "aggravation."[21] Fish also show signs of distress when they are taken out of the water and allowed to flap around in a net or on dry land until they die. Surely it is only because fish do not yelp or whimper in a way that we can hear that otherwise decent people can think it a pleasant way of spending an afternoon to sit by the water dangling a hook while previously caught fish die slowly beside them.

In 1976 the British Royal Society for the Prevention of Cruelty to Animals set up an independent panel of inquiry into shooting and angling. The panel was chaired by Lord Medway, a noted zoologist, and made up of experts outside the RSPCA. The inquiry examined in detail evidence on whether fish can feel pain, and concluded unequivocally that the evidence for pain in fish is as strong as the evidence for pain in other vertebrate animals.[22] People more concerned about causing pain than about killing may ask: Assuming fish *can* suffer, how much *do* they actually suffer in the normal process of commercial fishing? It may seem that fish, unlike birds and mammals, are not made to suffer in the process of rearing them for our tables, since they are usually not reared at all: human beings interfere with them only to catch and kill them. Actually this is not always true: fish farming—which is as intensive a form of factory farming as raising feedlot beef—is a rapidly growing industry. It began with freshwater fish like trout, but the Norwegians developed a technique for producing salmon in cages in the sea, and other countries are now using this method for a variety of marine fish. The potential welfare problems of farmed fish, such as stocking densities, the denial of the migratory urge, stress during handling, and so on have not even been investigated. But even with fish who are not farmed, the death of a commercially caught fish is much more drawn out than the death of, say, a chicken, since fish are simply hauled up into the air and left to die. Since their gills can extract oxygen from water but not from air, fish out of the water cannot breathe. The fish on sale in your supermarket may have died slowly, from suffocation. If it was a deep-sea fish, dragged to the surface by the net of a trawler, it may have died painfully from decompression.

When fish are caught rather than farmed, the ecological argument against eating intensively reared animals does not apply to fish. We do not waste grain or soybeans by feeding them to fish in the ocean. Yet there is a different ecological argument that counts against the extensive commercial fishing of the oceans now practiced, and this is that we are rapidly fishing out the oceans. In recent years fish catches have declined dramatically. Several once-abundant species of fish, such as the herrings of Northern Europe, the California sardines, and the New England haddock, are now so scarce as to be, for commercial purposes, extinct. Modern fishing fleets trawl the fishing grounds systematically with fine-gauge nets that catch everything in their way. The nontarget species—known in the industry as "trash"—may make up as much as half the catch.[23] Their bodies are thrown overboard. Because trawling involves dragging a huge net along the previously undisturbed bottom of the ocean, it damages the fragile ecology of the seabed. Like other ways of producing animal food, such fishing is also wasteful of fossil fuels, consuming more energy than it produces.[24] The nets used by the tuna fishing industry, moreover, also catch thousands of dolphins every year, trapping them underwater and drowning them. In addition to the disruption of ocean ecology caused by all this overfishing there are bad consequences for humans too. Throughout the world, small coastal villages that live by fishing are finding their traditional source of food and income drying up. From the communities on Ireland's west coast to the Burmese and Malayan fishing villages the story is the same. The fishing industry of the developed nations has become one more form of redistribution from the poor to the rich.

So out of concern for both fish and human beings we should avoid eating fish. Certainly those who continue to eat fish while refusing to eat other animals have taken a major step away from speciesism; but those who eat neither have gone one step further.

When we go beyond fish to the other forms of marine life commonly eaten by humans, we can no longer be quite so confident about the existence of a capacity for pain. Crustacea—lobster, crabs, prawns, shrimps—have nervous systems very different from our own. Nevertheless, Dr. John Baker, a zoologist at the University of Oxford and a fellow of the Royal Society, has stated that their sensory organs are highly developed, their nervous systems complex, their nerve cells very similar to our own, and their responses to certain stimuli immediate and vigorous. Dr. Baker therefore believes that lobster, for example, can feel pain. He is also clear that the standard method of killing lobster—dropping them into boiling water—can cause pain for as long as two minutes. He experimented with other methods sometimes said to be more humane, such as putting them in cold water and heating them slowly, or leaving them in fresh water until they cease to move, but found that both of these led to more prolonged struggling and, apparently, suffering.[25] If crustacea can suffer, there must be a great deal of suffering involved, not only in the method by which they are killed, but also in the ways in which they are transported and kept alive at markets. To keep them fresh they are frequently simply packed, alive, on top of each other. So even if there is some room for doubt about the capacity of these animals to feel pain, the fact that they may be suffering a great deal, combined with the absence of any need to eat them on our part, makes the verdict plain: they should receive the benefit of the doubt.

Oysters, clams, mussels, scallops, and the like are mollusks, and mollusks are in general very simple organisms. (There is an exception: the octopus is a mollusk, but far more developed, and presumably more sentient, than its distant mollusk relatives.) With creatures like oysters, doubts about a capacity for pain are considerable; and in the first edition of this book I suggested that somewhere between a shrimp and an oyster seems as good a place to draw the line as any. Accordingly, I continued occasionally to eat oysters, scallops, and mussels for some time after I became, in every other respect, a vegetarian. But while one cannot with any confidence say that these creatures do feel pain, so one can equally have little confidence in saying that they do not feel pain. Moreover, if they do feel pain, a meal of oysters or mussels would inflict pain on a considerable number of creatures. Since it is so easy to avoid eating them, I now think it better to do so.[26]

This takes us to the end of the evolutionary scale, so far as creatures we normally eat are concerned; essentially, we are left with a vegetarian diet. The traditional vegetarian diet, however, includes animal products, such as eggs and milk. Some have tried to accuse vegetarians of inconsistency here. "Vegetarian," they say, is a word that has the same root as "vegetable" and a vegetarian should eat only food of vegetable origin. Taken as a verbal quibble, this criticism is historically inaccurate. The term "vegetarian" came into general use as a result of the formation of the Vegetarian Society in England in 1847. Since the rules of the society permit the use of eggs and milk, the term "vegetarian" is properly applied to those who use these animal products. Recognizing this linguistic *fait accompli*, those who eat neither animal flesh nor eggs nor milk nor foods made from milk call themselves "vegans." The verbal point, however, is not the important one. What we should ask is whether the use of these other animal products is morally justifiable. This question is a real one because it is possible to be adequately nourished without consuming any animal products at all—a fact that is not widely known, although most people now know that vegetarians can live long and healthy lives. I shall say more on the topic of nutrition later in this chapter; for the present it is enough to know that we can do without eggs and milk. But is there any reason why we should?

We have seen that the egg industry is one of the most ruthlessly intensive forms of modern factory farming, exploiting hens relentlessly to produce the most eggs at the least cost. Our obligation to boycott this type of farming is as strong as our obligation to boycott intensively produced pork or chicken. But what of free-range eggs, assuming you can get them? Here the ethical objections are very much less. Hens provided with both shelter and an outdoor run to walk and scratch around in live comfortably. They do not appear to mind the removal of their eggs. The main grounds for objection are that the male chicks of the egg-laying strain will have been killed on hatching, and the hens themselves will be killed when they cease to lay productively. The question is, therefore, whether the pleasant lives of the hens (plus the benefits to us of the eggs) are sufficient to outweigh the killing that is a part of the system. One's answer to that will depend on one's view about killing, as distinct from the infliction of suffering. There is some further discussion of the relevant philosophical issues in the final chapter of this book.[27] In keeping with the reasons given there, I do not, on balance, object to free-range egg production.

Milk and milk products like cheese and yogurt raise different issues. We have seen in Chapter 3 that dairy production can be distressing for the cows and their calves in several ways: the necessity of making the cow pregnant, and the subsequent separation of the cow and her calf; the increasing degree of confinement on many farms; the health and stress problems caused by feeding cows very rich diets and breeding them for ever-greater milk yields; and now the prospect of further stress from daily injections of bovine growth hormone.

In principle, there is no problem in doing without dairy products. Indeed, in many parts of Asia and Africa, the only milk ever consumed is human milk, for infants. Many adults from these parts of the world lack the ability to digest the lactose that milk contains, and they become ill if they drink milk. The Chinese and Japanese have long used soybeans to make many of the things we make from dairy products. Soy milks are now widely available in Western countries, and tofu ice cream is popular with those trying to reduce their intake of fat and cholesterol. There are even cheeses, spreads, and yogurts made from soybeans.

Vegans, then, are right to say that we ought not to use dairy products. They are living demonstrations of the practicality and nutritional soundness of a diet that is totally free from the exploitation of other animals. At the same time, it should be said that, in our present speciesist world, it is not easy to keep so strictly to what is morally right. A reasonable and defensible plan of action is to change your diet at a measured pace with which you can feel comfortable. Although in principle all dairy products are replaceable, in practice in Western societies it is much more difficult to cut out meat and dairy products than it is to elminate meat alone. Until you start reading food labels with an eye to avoiding dairy products, you will never believe how many foods contain them. Even buying a tomato sandwich becomes a problem, since it will probably be spread either with butter, or with a margarine containing whey or nonfat milk. There is little gained for animals if you give up animal flesh and battery eggs, and simply replace them with an increased amount of cheese. On the other hand, the following is, if not ideal, a reasonable and practical strategy:

- replace animal flesh with plant foods;
- replace factory farm eggs with free-range eggs if you can get them; otherwise avoid eggs;
- replace the milk and cheese you buy with soymilk, tofu, or other plant foods, but do not feel obliged to go to great lengths to avoid all food containing milk products.

Eliminating speciesism from one's dietary habits is very difficult to do all at once. People who adopt the strategy I support here have made a clear public commitment to the movement against animal exploitation. The most urgent task of the Animal Liberation movement is to persuade as many people as possible to make this commitment, so that the boycott will spread and gain attention. If because of an admirable desire to stop all forms of exploitation of animals immediately we convey the impression that unless one gives up milk products one is no better than those who still eat animal flesh, the result may be that many people are deterred from doing anything at all, and the exploitation of animals will continue as before.

These, at least, are some of the answers to problems that are likely to face nonspeciesists who ask what they should and should not eat. As I said at the beginning of this section, my remarks are intended to be no more than suggestions. Sincere nonspeciesists may well disagree among themselves about the details. So long as there is agreement on the fundamentals this should not disrupt efforts toward a common goal.

Many people are willing to admit that the case for vegetarianism is strong. Too often, though, there is a gap between intellectual conviction and the action needed to break a lifetime habit. There is no way in which books can bridge this gap; ultimately it is up to each one of us to put our convictions into practice. But I can try, in the next few pages, to narrow the gap. My aim is to make the transition from an omnivorous diet to a vegetarian one much easier and more attractive, so that instead of seeing the change of diet as an unpleasant duty the reader looks forward to a new and interesting cuisine, full of fresh foods as well as unusual meatless dishes from Europe, China, and the Middle East, dishes so varied as to make the habitual meat, meat, and more meat of most Western diets stale and repetitive by comparison. The enjoyment of such a cuisine is enhanced by the knowledge that its good taste and nourishing qualities were provided directly by the earth, neither wasting what the earth produces, nor requiring the suffering and death of any sentient being.

Vegetarianism brings with it a new relationship to food, plants, and nature. Flesh taints our meals. Disguise it as we may, the fact remains that the centerpiece of our dinner has come to us from the slaughterhouse, dripping blood. Untreated and unrefrigerated, it soon begins to putrefy and stink. When we eat it, it sits heavily in our stomachs, blocking our digestive processes until, days later, we struggle to excrete it.[28] When we eat plants, food takes on a different quality. We take from the earth food that is ready for us and does not fight against us as we take it. Without meat to deaden the palate we experience an extra delight in fresh vegetables taken straight from the ground. Personally, I found the idea of picking my own dinner so satisfying that shortly after becoming a vegetarian I began digging up part of our backyard and growing some of my own vegetables—something that I had never thought of doing previously, but that several of my vegetarian friends were also doing. In this way dropping flesh-meat from my diet brought me into closer contact with plants, the soil, and the seasons.

Cooking, too, was something I became interested in only after I became a vegetarian. For those brought up on the usual Anglo-Saxon menus, in which the main dish consists of meat supplemented by two overcooked vegetables, the elimination of meat poses an interesting challenge to the imagination. When I speak in public about the issues discussed in this book, I am often asked about what one can eat instead of meat, and it is clear from the way the question is phrased that the questioner has mentally subtracted the chop or hamburger from his or her plate, leaving the mashed potatoes and boiled cabbage, and is wondering what to put in place of the meat. A heap of soybeans perhaps?

There may be those who would enjoy such a meal, but for most tastes the answer is to rethink the entire idea of the main course, so that it consists of a combination of ingredients, perhaps with a salad on the side, instead of detached items. Good Chinese dishes, for instance, are superb combinations of one or more high-protein ingredients—in vegetarian Chinese cooking, they may include tofu, nuts, bean sprouts, mushrooms, or wheat gluten, with fresh, lightly cooked vegetables and rice. An Indian curry using lentils for protein, served over brown rice with some fresh sliced cucumber for light relief, makes an equally satisfying meal, as does an Italian vegetarian lasagna with salad. You can even make "tofu meatballs" to put on top of your spaghetti. A simpler meal might consist of whole grains and vegetables. Most Westerners eat very little millet, whole wheat, or buckwheat, but these grains can form the basis of a dish that is a refreshing change. In the first edition of this book I provided some recipes and hints on vegetarian cooking to help readers make the transition to what was, then, still an unusual diet; but in the intervening years so many excellent vegetarian cookbooks have been published that the assistance I was able to provide seems quite unnecessary now. (I've recommended a few cookbooks in Appendix 2.) Some people find it hard, at first, to change their attitude to a meal. Getting used to meals without a central piece of animal flesh may take time, but once it has happened you will have so many interesting new dishes to choose from that you will wonder why you ever thought it would be difficult to do without flesh foods.

Apart from the tastiness of their meals, people contemplating vegetarianism are most likely to worry about whether they will be adequately nourished. These worries are entirely groundless. Many parts of the world have vegetarian cultures whose members have been as healthy, and often healthier, than nonvegetarians living in similar areas. Strict Hindus have been vegetarians for more than two thousand years. Gandhi, a lifelong vegetarian, was close to eighty when an assassin's bullet ended his active life. In Britain, where there has now been an official vegetarian movement for more than 140 years, there are third- and fourth-generation vegetarians. Many prominent vegetarians, such as Leonardo da Vinci, Leo Tolstoy, and George Bernard Shaw, have lived long, immensely creative lives. Indeed, most people who have reached exceptional old age have eaten little or no meat. The inhabitants of the Vilcabamba valley in Ecuador frequently live to be more than one hundred years old, and men as old as 123 and 142 years have been found by scientists; these people eat less than one ounce of meat a week. A study of all living centenarians in Hungary found that they were largely vegetarian.[29]

That meat is unnecessary for physical endurance is shown by a long list of successful athletes who do not eat it, a list that includes Olympic long-distance swimming champion Murray Rose, the famous Finnish distance runner Paavo Nurmi, basketball star Bill Walton, the "ironman" triathlete Dave Scott, and 400-meter Olympic hurdle champion Edwin Moses.

Many vegetarians claim that they feel fitter, healthier, and more zestful than when they ate meat. A great deal of new evidence now supports them. The 1988 United States Surgeon General's Report on Nutrition and Health cites a major study indicating that the death rate for heart attacks of vegetarians between the ages of thirty-five and sixty-four is only 28 percent of the rate for Americans in general in that age group. For older vegetarians the rate of death from heart attacks was still less than half that of nonvegetarians. The same study showed that vegetarians who ate eggs and dairy products had cholesterol levels 16 percent lower than those of meat eaters, and vegans had cholesterol levels 29 percent lower. The report's main recommendations were to reduce consumption of cholesterol and fat (especially saturated fat), and increase consumption of whole grain foods and cereal products, vegetables (including dried beans and peas) and fruits. A recommendation to reduce cholesterol and saturated fat is, in effect, a recommendation to avoid meat (except perhaps chicken from which the skin has been removed), and cream, butter, and all except low-fat dairy products.[30] The report was widely criticized for failing to be more specific in saying this—a vagueness due, apparently, to successful lobbying by groups like the National Cattlemen's Association and the Dairy Board.[31] Whatever lobbying took place failed however, to prevent the section on cancer from reporting that studies have found an association between breast cancer and meat intake, and also between eating meat, especially beef, and cancer of the large bowel. The American Heart Association has also been recommending, for many years, that Americans reduce their meat intake.[32] Diets designed for health and longevity like the Pritikin plan and the McDougall plan are either largely or entirely vegetarian.[33]

Nutritional experts no longer dispute about whether animal flesh is essential; they now agree that it is not. If ordinary people still have misgivings about doing without it, these misgivings are based on ignorance. Most often this ignorance is about the nature of protein. We are frequently told that protein is an important element in a sound diet, and that meat is high in protein. Both these statements are true, but there are two other things that we are told less often. The first is that the average American eats too much protein. The protein intake of the average American exceeds the generous level recommended by the National Academy of Sciences by 45 percent. Other estimates say that most Americans consume between two and four times as much meat as the body can use. Excess protein cannot be stored. Some of it is excreted, and some may be converted by the body to carbohydrate, which is an expensive way to increase one's carbohydrate intake.[34]

The second thing to know about protein is that meat is only one among a great variety of foods containing protein, its chief distinction being that it is the most expensive. It was once thought that meat protein was of superior quality, but as long ago as 1950 the British Medical Association's Committee on Nutrition stated:

> It is generally accepted that it is immaterial whether the essential protein units are derived from plant or animal foods, provided that they supply an appropriate mixture of the units in assimilable form.[35]

More recent research has provided further confirmation of this conclusion. We now know that the nutritional value of protein consists in the essential amino acids it contains, since these determine how much of the protein the body can use. While it is true that animal foods, especially eggs and milk, have a very well-balanced amino acid composition, plant foods like soybeans and nuts also contain a broad range of these nutrients. Moreover by eating different kinds of plant proteins at the same time it is easy to put together a meal that provides protein entirely equivalent to that of animal protein. This principle is called "protein complementarity," but you do not need to know much about nutrition to apply it. The peasant who eats his beans or lentils with rice or corn is practicing protein complementarity. So is the mother who gives her child a peanut butter sandwich on whole wheat bread—a combination of peanuts and wheat, both of which contain protein. The different forms of protein in the different foods combine with each other in such a way that the body absorbs more protein if they are eaten together than if they were eaten separately. Even without the complementary effect of combining different proteins, however, most of the plant foods we eat—not just nuts, peas, and beans, but even wheat, rice, and potatoes—contain enough protein in themselves to provide our bodies with the protein we need. If we avoid junk foods that are high in sugar or fats and nothing else, about the only way we can fail to get enough protein is if we are on a diet that is insufficient in calories.[36]

Protein is not the only nutrient in meat, but the others can all easily be obtained from a vegetarian diet without special care. Only vegans, who take no animal products at all, need to be especially careful about their diet. There appears to be one, and only one, necessary nutrient that is not normally available from plant sources, and this is vitamin B12, which is present in eggs and milk, but not in a readily assimilable form in plant foods. It can, however, be obtained from seaweeds such as kelp, from a soy sauce made by the traditional Japanese fermentation method, or from tempeh, a fermented soybean product eaten in parts of Asia, and often now available in health food stores in the West. It is also possible that it is produced by microorganisms in our own intestines. Studies of vegans who have not taken any apparent source of B12 for many years have shown their blood levels of this vitamin still to be within the normal range. Nevetheless to make sure of avoiding a deficiency, it is simple and inexpensive to take vitamin tablets containing B12. The B12 in these tablets is obtained from bacteria grown on plant foods. Studies of children in vegan families have shown that they develop normally on diets that contain a B12 supplement but no animal food after weaning.[37]

I have tried in this chapter to answer the doubts about becoming a vegetarian that can easily be articulated and expressed. But some people have a deeper resistance that makes them hesitate. Perhaps the reason for hesitation is a fear of being thought a crank by one's friends. When my wife and I began to think about becoming vegetarians we talked about this. We worried that we would be cutting ourselves off from our nonvegetarian friends and at that time none of our long-established friends was vegetarian. The fact that we became vegetarians together certainly made the decision easier for both of us, but as things turned out we need not have worried. We explained our decision to our friends and they saw that we had good reasons for it. They did not all become vegetarians, but they did not cease to be our friends either; in fact I think they rather enjoyed inviting us to dinner and showing us how well they could cook without meat. Of course, it is possible that you will encounter people who consider you a crank. This is much less likely now than it was a few years ago, because there are so many more vegetarians. But if it should happen, remember that you are in good company. All the best reformers—those who first opposed the slave trade, nationalistic wars, and the exploitation of children working a fourteen-hour day in the factories of the Industrial Revolution—were at first derided as cranks by those who had an interest in the abuses they were opposing.

Chapter 5

Man's Dominion . . .

a short history of speciesism

To end tyranny we must first understand it. As a practical matter, the rule of the human animal over other animals expresses itself in the manner we have seen in Chapters 2 and 3, and in related practices like the slaughter of wild animals for sport or for their furs. These practices should not be seen as isolated aberrations. They can be properly understood only as the manifestations of the ideology of our species—that is, the attitudes which we, as the dominant animal, have toward the other animals.

In this chapter we shall see how, at different periods, outstanding Western thinkers formulated and defended the attitudes to animals that we have inherited. I concentrate on the "West" not because other cultures are inferior—the reverse is true, so far as attitudes to animals are concerned—but because Western ideas have, over the past two or three centuries, spread out from Europe until today they set the mode of thought for most human societies, whether capitalist or communist.

Though the material that follows is historical, my aim in presenting it is not. When an attitude is so deeply ingrained in our thought that we take it as an unquestioned truth, a serious and consistent challenge to that attitude runs the risk of ridicule. It may be possible to shatter the complacency with which the attitude is held by a frontal attack. This is what I have tried to do in the preceding chapters. An alternative strategy is to attempt to undermine the plausibility of the prevailing attitude by revealing its historical origins.

The attitudes toward animals of previous generations are no longer convincing because they draw on presuppositions—religious, moral, metaphysical—that are now obsolete. Because we do not defend our attitudes to animals in the way that Saint

Thomas Aquinas, for example, defended his attitudes to animals, we may be ready to accept that Aquinas used the religious, moral, and metaphysical ideas of his time to mask the naked self-interest of human dealings with other animals. If we can see that past generations accepted as right and natural attitudes that we recognize as ideological camouflages for self-serving practices—and if, at the same time, it cannot be denied that we continue to use animals to further our own minor interests in violation of their major interests—we may be persuaded to take a more skeptical view of those justifications of our own practices that we ourselves have taken to be right and natural.

Western attitudes to animals have roots in two traditions: Judaism and Greek antiquity. These roots unite in Christianity, and it is through Christianity that they came to prevail in Europe. A more enlightened view of our relations with animals emerges only gradually, as thinkers begin to take positions that are relatively independent of the church; and in fundamental respects we still have not broken free of the attitudes that were unquestioningly accepted in Europe until the eighteenth century. We may divide our historical discussion, therefore, into three parts: pre-Christian, Christian, and the Enlightenment and after.

Pre-Christian Thought

The creation of the universe seems a fit starting point. The biblical story of the creation sets out very clearly the nature of the relationship between man and animal as the Hebrew people conceived it to be. It is a superb example of myth echoing reality:

> And God said, Let the earth bring forth the living creature after his kind, cattle and creeping thing, and beast of the earth after his kind: and it was so.
>
> And God made the beast of the earth after his kind, and cattle after their kind, and every thing that creepeth upon the earth after his kind: and God saw that it was good.
>
> And God said, Let us make man in our image, after our likeness: and let them have dominion over the fish of the sea, and over the fowl of the air, and over the earth, and over every creeping thing that creepeth upon the earth.

> So God created man in his own image, in the image of God created he him; male and female created he them.
>
> And God blessed them, and God said unto them, Be fruitful, and multiply, and replenish the earth, and subdue it; and have dominion over the fish of the sea, and over the fowl of the air, and over every living thing that moveth upon the earth.[1]

The Bible tells us that God made man in His own image. We may regard this as man making God in his own image. Either way, it allots human beings a special position in the universe, as beings that, alone of all living things, are God-like. Moreover, God is explicitly said to have given man dominion over every living thing. It is true that, in the Garden of Eden, this dominion may not have involved killing other animals for food. Genesis 1:29 suggests that at first human beings lived off the herbs and fruits of the trees, and Eden has often been pictured as a scene of perfect peace, in which killing of any kind would have been out of place. Man ruled, but in this earthly paradise his was a benevolent despotism.

After the fall of man (for which the Bible holds a woman and an animal responsible), killing animals clearly was permissible. God himself clothed Adam and Eve in animal skins before driving them out of the Garden of Eden. Their son Abel was a keeper of sheep and made offerings of his flock to the Lord. Then came the flood, when the rest of creation was nearly wiped out to punish man for his wickedness. When the waters subsided Noah thanked God by making burnt offerings "of every clean beast, and of every clean fowl." In return, God blessed Noah and gave the final seal to man's dominion:

> And God blessed Noah and his sons, and said unto them, Be fruitful, and multiply, and replenish the earth.
>
> And the fear of you and the dread of you shall be upon every beast of the earth, and upon every fowl of the air, upon all that moveth upon the earth, and upon all the fishes of the sea; into your hands are they delivered.
>
> Every moving thing that liveth shall be meat for you; even as the green herb have I given you all things.[2]

This is the basic position of the ancient Hebrew writings toward nonhumans. There is again an intriguing hint that in the original state of innocence we were vegetarian, eating only "the green herb," but that after the fall, the wickedness that followed it, and the flood, we were given permission to add animals to our diet. Beneath the assumption of human dominion that this permission implies, a more compassionate vein of thought still occasionally emerges. The prophet Isaiah condemned animal sacrifices, and the book of Isaiah contains a lovely vision of the time when the wolf will dwell with the lamb, the lion will eat straw like the ox, and "they shall not hurt nor destroy in all my holy mountain." This, however, is a utopian vision, not a command to be followed immediately. Other scattered passages in the Old Testament encourage some degree of kindliness toward animals, so that it is possible to argue that wanton cruelty was prohibited, and that "dominion" is really more like a "stewardship," in which we are responsible to God for the care and well-being of those placed under our rule. Nevertheless there is no serious challenge to the overall view, laid down in Genesis, that the human species is the pinnacle of creation and has God's permission to kill and eat other animals.

The second ancient tradition of Western thought is that of Greece. Here we find, at first, conflicting tendencies. Greek thought was not uniform, but divided into rival schools, each taking its basic doctrines from some great founder. One of these, Pythagoras, was a vegetarian and encouraged his followers to treat animals with respect, apparently because he believed that the souls of dead men migrated to animals. But the most important school was that of Plato and his pupil, Aristotle.

Aristotle's support for slavery is well known; he thought that some men are slaves by nature and that slavery is both right and expedient for them. I mention this not in order to discredit Aristotle, but because it is essential for understanding his attitude to animals. Aristotle holds that animals exist to serve the purposes of human beings, although, unlike the author of Genesis, he does not drive any deep gulf between human beings and the rest of the animal world.

Aristotle does not deny that man is an animal; in fact he defines man as a rational animal. Sharing a common animal nature, however, is not enough to justify equal consideration. For Aristotle the man who is by nature a slave is undoubtedly a human being, and is as capable of feeling pleasure and pain as any other human being; yet because he is supposed to be inferior to the free man in his reasoning powers, Aristotle regards him as a "living instrument." Quite openly, Aristotle juxtaposes the two elements in a single sentence: the slave is one who "though remaining a human being, is also an article of property."[3]

If the difference in reasoning powers between human beings is enough to make some masters and others their property, Aristotle must have thought the rights of human beings to rule over other animals too obvious to require much argument. Nature, he held, is essentially a hierarchy in which those with less reasoning ability exist for the sake of those with more:

> Plants exist for the sake of animals, and brute beasts for the sake of man—domestic animals for his use and food, wild ones (or at any rate most of them) for food and other accessories of life, such as clothing and various tools.
>
> Since nature makes nothing purposeless or in vain, it is undeniably true that she has made all animals for the sake of man.[4]

It was the views of Aristotle, rather than those of Pythagoras, that were to become part of the later Western tradition.

Christian Thought

Christianity was in time to unite Jewish and Greek ideas about animals. But Christianity was founded and became powerful under the Roman Empire, and we can see its initial effect best if we compare Christian attitudes with those they replaced.

The Roman Empire was built by wars of conquest, and needed to devote much of its energy and revenue to the military forces that defended and extended its vast territory. These conditions did not foster sentiments of sympathy for the weak. The martial virtues set the tone of the society. Within Rome itself, far from the fighting on the frontiers, the character of Roman citizens was supposedly toughened by the so-called games. Although every schoolboy knows how Christians were thrown to the lions in the

Colosseum, the significance of the games as an indication of the possible limits of sympathy and compassion of apparently—and in other respects genuinely—civilized people is rarely appreciated. Men and women looked upon the slaughter of both human beings and other animals as a normal source of entertainment; and this continued for centuries with scarcely a protest.

The nineteenth-century historian W. E. H. Lecky gives the following account of the development of the Roman games from their beginning as a combat between two gladiators:

> The simple combat became at last insipid, and every variety of atrocity was devised to stimulate the flagging interest. At one time a bear and a bull, chained together, rolled in fierce combat across the sand; at another, criminals dressed in the skins of wild beasts were thrown to bulls, which were maddened by red-hot irons, or by darts tipped with burning pitch. Four hundred bears were killed on a single day under Caligula.... Under Nero, four hundred tigers fought with bulls and elephants. In a single day, at the dedication of the Colosseum by Titus, five thousand animals perished. Under Trajan, the games continued for one hundred and twenty-three successive days. Lions, tigers, elephants, rhinoceroses, hippopotami, giraffes, bulls, stags, even crocodiles and serpents were employed to give novelty to the spectacle. Nor was any form of human suffering wanting.... Ten thousand men fought during the games of Trajan. Nero illumined his gardens during the night by Christians burning in their pitchy shirts. Under Domitian, an army of feeble dwarfs was compelled to fight.... So intense was the craving for blood, that a prince was less unpopular if he neglected the distribution of corn than if he neglected the games.[5]

The Romans were not without any moral feelings. They showed a high regard for justice, public duty, and even kindness to others. What the games show, with hideous clarity, is that there was a sharp limit to these moral feelings. If a being came within this limit, activities comparable to what occurred at the games would have been an intolerable outrage; when a being was outside the sphere of moral concern, however, the infliction of suffering was merely entertaining. Some human beings—criminals and military captives especially—and all animals fell outside this sphere.

It is against this background that the impact of Christianity must be assessed. Christianity brought into the Roman world the idea of the uniqueness of the human species, which it inherited from the Jewish tradition but insisted upon with still greater emphasis because of the importance it placed on the human being's immortal soul. Human beings, alone of all beings living on earth, were destined for life after bodily death. With this came the distinctively Christian idea of the sanctity of all human life.

There have been religions, especially in the East, which have taught that all life is sacred; and there have been many others that have held it gravely wrong to kill members of one's own social, religious, or ethnic group; but Christianity spread the idea that every human life—and only human life—is sacred. Even the newborn infant and the fetus in the womb have immortal souls, and so their lives are as sacred as those of adults.

In its application to human beings, the new doctrine was in many ways progressive, and led to an enormous expansion of the limited moral sphere of the Romans; so far as other species are concerned, however, this same doctrine served to confirm and further depress the lowly position nonhumans had in the Old Testament. While it asserted human dominion over other species, the Old Testament did at least show flickers of concern for their sufferings. The New Testament is completely lacking in any injunction against cruelty to animals, or any recommendation to consider their interests. Jesus himself is described as showing apparent indifference to the fate of nonhumans when he induced two thousand swine to hurl themselves into the sea—an act which was apparently quite unnecessary, since Jesus was well able to cast out devils without inflicting them upon any other creature.[6] Saint Paul insisted on reinterpreting the old Mosaic law that forbade muzzling the ox that trod out the corn: "Doth God care for oxen?" Paul asks scornfully. No, he answered, the law was intended "altogether for our sakes."[7]

The example given by Jesus was not lost on later Christians. Referring to the incident of the swine and the episode in which Jesus cursed a fig tree, Saint Augustine wrote:

> Christ himself shows that to refrain from the killing of animals and the destroying of plants is the height of superstition, for judging that there are no common rights between us and the beasts and trees, he sent the devils into a herd of swine and with a curse withered the tree on which he found no fruit.... Surely the swine had not sinned, nor had the tree.

Jesus was, according to Augustine, trying to show us that we need not govern our behavior toward animals by the moral rules that govern our behavior toward humans. That is why he transferred the devils to swine instead of destroying them as he could easily have done.[8]

On this basis the outcome of the interaction of Christian and Roman attitudes is not difficult to guess. It can be seen most clearly by looking at what happened to the Roman games after the conversion of the empire to Christianity. Christian teaching was implacably opposed to gladiatorial combats. The gladiator who survived by killing his opponent was regarded as a murderer. Mere attendance at these combats made the Christian liable to excommunication, and by the end of the fourth century combats between human beings had been suppressed altogether. On the other hand, the moral status of killing or torturing any nonhuman remained unchanged. Combats with wild animals continued into the Christian era, and apparently declined only because the declining wealth and extent of the empire made wild animals more difficult to obtain. Indeed, these combats may still be seen, in the modern form of the bullfight, in Spain and Latin America.

What is true of the Roman games is also true more generally. Christianity left nonhumans as decidedly outside the pale of sympathy as they ever were in Roman times. Consequently, while attitudes to human beings were softened and improved beyond recognition, attitudes to other animals remained as callous and brutal as they were in Roman times. Indeed, not only did Christianity fail to temper the worst of Roman attitudes toward other animals; it unfortunately succeeded in extinguishing for a long, long time the spark of a wider compassion that had been kept alight by a tiny number of more gentle people.

There had been just a few Romans who had shown compassion for suffering, whatever the being who suffered, and repulsion at the use of sentient creatures for human pleasure, whether at the gourmet's table or in the arena. Ovid, Seneca, Porphyry, and Plutarch all wrote along these lines, Plutarch having the honor, according to Lecky, of being the first to advocate strongly the kind treatment of animals on the ground of universal benevolence, independently of any belief in the transmigration of souls.[9] We have to wait nearly sixteen hundred years, however, before any Christian writer attacks cruelty to animals with similar emphasis and detail on any ground other than that it may encourage a tendency toward cruelty to humans.

A few Christians expressed some concern for animals. There is a prayer written by Saint Basil that urges kindness to animals, a remark by Saint John Chrysostom to the same effect, and a teaching of Saint Isaac the Syrian. There were even some saints who, like Saint Neot, sabotaged hunts by rescuing stags and hares from the hunters.[10] But these figures failed to divert mainstream Christian thinking from its exclusively speciesist preoccupation. To demonstrate this lack of influence, instead of tracing the development of Christian views on animals through the early Church Fathers to the medieval scholastics—a tedious process, since there is more repetition than development—it will be better to consider in more detail than would otherwise be possible the position of Saint Thomas Aquinas.

Aquinas's enormous *Summa Theologica* was an attempt to grasp the sum of theological knowledge and reconcile it with the worldly wisdom of the philosophers, though for Aquinas, Aristotle was so preeminent in his field that he is referred to simply as "the Philosopher." If any single writer may be taken as representative of Christian philosophy prior to the Reformation, and of Roman Catholic philosophy to this day, it is Aquinas.

We may begin by asking whether, according to Aquinas, the Christian prohibition on killing applies to creatures other than humans, and if not, why not. Aquinas answers:

> There is no sin in using a thing for the purpose for which it is. Now the order of things is such that the imperfect are for the perfect.... Things, like plants which merely have life, are all alike for animals, and all animals are for man. Wherefore it is not unlawful if men use plants for the good of animals,

> and animals for the good of man, as the Philosopher states (*Politics* I, 3).
>
> Now the most necessary use would seem to consist in the fact that animals use plants, and men use animals, for food, and this cannot be done unless these be deprived of life, wherefore it is lawful both to take life from plants for the use of animals, and from animals for the use of men. In fact this is in keeping with the commandment of God himself (*Genesis* i, 29, 30 and *Genesis* ix, 3).[11]

For Aquinas the point is not that killing for food is in itself necessary and therefore justifiable (since Aquinas knew of sects like the Manichees in which the killing of animals was forbidden, he could not have been entirely ignorant of the fact that human beings can live without killing animals, but we shall overlook this for the moment); it is only the "more perfect" who are entitled to kill for this reason. Animals that kill human beings for food are in a quite different category:

> Savagery and brutality take their names from a likeness to wild beasts. For animals of this kind attack man that they may feed on his body, and not for some motive of justice, the consideration of which belongs to reason alone.[12]

Human beings, of course, would not kill for food unless they had first considered the justice of so doing!

So human beings may kill other animals and use them for food; but are there perhaps other things that we may not do to them? Is the suffering of other creatures in itself an evil? If so would it not for that reason be wrong to make them suffer, or at least to make them suffer unnecessarily?

Aquinas does not say that cruelty to "irrational animals" is wrong in itself. He has no room for wrongs of this kind in his moral schema, for he divides sins into those against God, those against oneself, and those against one's neighbor. So the limits of morality once again exclude nonhumans. There is no category for sins against them.[13]

Perhaps although it is not a sin to be cruel to nonhumans, it is charitable to be kind to them? No, Aquinas explicitly excludes this possibility as well. Charity, he says, does not extend to irrational creatures for three reasons: they are "not competent, properly speaking, to possess good, this being proper to rational creatures"; we have no fellow-feeling with them; and, finally, "charity is based on the fellowship of everlasting happiness, to which the irrational creature cannot attain." It is only possible to love these creatures, we are told, "if we regard them as the good things that we desire for others," that is, "to God's honor and man's use." In other words, we cannot lovingly give food to turkeys because they are hungry, but only if we think of them as someone's Christmas dinner.[14]

All this might lead us to suspect that Aquinas simply doesn't believe that animals other than human beings are capable of suffering at all. This view has been held by other philosophers and, for all its apparent absurdity, to attribute it to Aquinas would at least excuse him of the charge of indifference to suffering. This interpretation, however, is ruled out by his own words. In the course of a discussion of some of the mild injunctions against cruelty to animals in the Old Testament, Aquinas proposes that we distinguish reason and passion. So far as reason is concerned, he tells us:

> It matters not how man behaves to animals, because God has subjected all things to man's power and it is in this sense that the Apostle says that God has no care for oxen, because God does not ask of man what he does with oxen or other animals.

On the other hand, where passion is concerned, our pity is aroused by animals, because "even irrational animals are sensible to pain"; nevertheless, Aquinas regards the pain that animals suffer as insufficient reason to justify the Old Testament injunctions, and therefore adds:

> Now it is evident that if a man practice a pitiable affection for animals, he is all the more disposed to take pity on his fellow-men, wherefore it is written (*Proverbs* xii, 10) "The just regardeth the life of his beast."[15]

So Aquinas arrives at the often to be repeated view that the only reason against cruelty to animals is that it may lead to cruelty to human beings. No argument could reveal the essence of speciesism more clearly.

Aquinas's influence has lasted. As late as the middle of the nineteenth century, Pope Pius IX refused to allow a Society for the Prevention of Cruelty to Animals to be established in Rome, on the grounds that to do so would imply that human beings have duties toward animals.[16] And we can bring this account right up to the second half of the twentieth century without finding significant modifications in the official position of the Roman Catholic Church. The following passage, from an American Roman Catholic text, makes an instructive comparison with the passage written seven hundred years ago, and quoted above, from Aquinas:

> In the order of nature, the imperfect is for the sake of the perfect, the irrational is to serve the rational. Man, as a rational animal, is permitted to use things below him in this order of nature for his proper needs. He needs to eat plants and animals to maintain his life and strength. To eat plants and animals, they must be killed. So killing is not, of itself, an immoral or unjust act.[17]

The point to notice about this text is that the author sticks so closely to Aquinas that he even repeats the assertion that it is necessary for human beings to eat plants *and* animals. The ignorance of Aquinas in this respect was surprising, but excusable given the state of scientific knowledge in his time; that a modern author, who would only need to look up a standard work on nutrition or take note of the existence of healthy vegetarians, should carry on the same error is incredible.

It was only in 1988 that an authoritative statement from the Roman Catholic Church indicated that the environmental movement is beginning to affect Catholic teachings. In his encyclical *Solicitudo Rei Socialis* ("On Social Concerns"), Pope John Paul II urged that human development should include "respect for the beings which constitute the natural world" and added:

> The dominion granted to man by the Creator is not an absolute power, nor can one speak of a freedom to "use and misuse," or to dispose of things as one pleases.... When it

> comes to the natural world, we are subject not only to biological laws, but also to moral ones, which cannot be violated with impunity.[18]

That a Pope should so clearly reject the absolute dominion view is very promising, but it is too early to say if it signals a historic and much-needed change of direction in Catholic teaching about animals and the environment.

There have, of course, been many humane Catholics who have done their best to ameliorate the position of their church with regard to animals, and they have had occasional successes. By stressing the degrading tendency of cruelty, some Catholic writers have felt themselves able to condemn the worst of human practices toward other animals. Yet most remain limited by the basic outlook of their religion. The case of Saint Francis of Assisi illustrates this.

Saint Francis is the outstanding exception to the rule that Catholicism discourages concern for the welfare of nonhuman beings. "If I could only be presented to the emperor," he is reported as saying, "I would pray him, for the love of God, and of me, to issue an edict prohibiting anyone from catching or imprisoning my sisters the larks, and ordering that all who have oxen or asses should at Christmas feed them particularly well." Many legends tell of his compassion, and the story of how he preached to the birds certainly seems to imply that the gap between them and humans was less than other Christians supposed.

But a misleading impression of the views of Saint Francis may be gained if one looks only at his attitude to larks and the other animals. It was not only sentient creatures whom Saint Francis addressed as his sisters: the sun, the moon, wind, fire, all were brothers and sisters to him. His contemporaries described him as taking "inward and outward delight in almost every creature, and when he handled or looked at them his spirit seemed to be in heaven rather than on earth." This delight extended to water, rocks, flowers, and trees. This is a description of a person in a state of religious ecstasy, deeply moved by a feeling of oneness with all of nature. People from a variety of religious and mystical traditions appear to have had such experiences, and have expressed similar feelings of universal love. Seeing Francis in this light makes the breadth of his love and compassion more readily comprehensible. It also enables us to see how his love for all creatures could coexist with a theological position that was quite orthodox in its speciesism. Saint Francis affirmed that "every creature proclaims: 'God made me for your sake, O man!'" The sun itself, he thought, shines for man. These beliefs were part of a cosmology that he never questioned; the force of his love for all creation, however, was not to be bound by such considerations.

While this kind of ecstatic universal love can be a wonderful source of compassion and goodness, the lack of rational reflection can also do much to counteract its beneficial consequences. If we love rocks, trees, plants, larks, and oxen equally, we may lose sight of the essential differences between them, most importantly, the differences in degree of sentience. We may then think that since we have to eat to survive, and since we cannot eat without killing something we love, it does not matter which we kill. Possibly it was for this reason that Saint Francis's love for birds and oxen appears not to have led him to cease eating them; and when he drew up the rules for the conduct of the friars in the order he founded, he gave no instruction that they were to abstain from meat, except on certain fast days.[19]

It may seem that the period of the Renaissance, with the rise of humanist thought in opposition to medieval scholasticism, would have shattered the medieval picture of the universe and brought down with it earlier ideas about the status of humans vis-à-vis the other animals. But Renaissance humanism was, after all, *humanism*; and the meaning of this term has nothing to do with humanitarianism, the tendency to act humanely.

The central feature of Renaissance humanism is its insistence on the value and dignity of human beings, and on the central place of human beings in the universe. "Man is the measure of all things," a phrase revived in Renaissance times from the ancient Greeks, is the theme of the period. Instead of a somewhat depressing concentration on original sin and the weakness of human beings in comparison to the infinite power of God, the Renaissance humanists emphasized the uniqueness of human beings, their free will, their potential, and their dignity; and they contrasted all this with the limited nature of the "lower animals."

Like the original Christian insistence on the sanctity of human life, this was in some ways a valuable advance in attitudes to human beings, but it left nonhumans as far below humans as they had ever been.

So the Renaissance writers wrote self-indulgent essays in which they said that "nothing in the world can be found that is more worthy of admiration than man"[20] and described humans as "the center of nature, the middle of the universe, the chain of the world."[21] If the Renaissance marks in some respect the beginning of modern thought, so far as attitudes to animals were concerned earlier modes of thought still maintained their hold.

Around this time, however, we may notice the first genuine dissenters: Leonardo da Vinci was teased by his friends for being so concerned about the sufferings of animals that he became a vegetarian;[22] and Giordano Bruno, influenced by the new Copernican astronomy, which allowed for the possibility of other planets, some of which could be inhabited, ventured to assert that "man is no more than an ant in the presence of the infinite." Bruno was burned at the stake in 1600 for refusing to recant his heresies.

Michel de Montaigne's favorite author was Plutarch, and his attack on the humanist assumptions of his age would have met with the approval of that gentle Roman:

> Presumption is our natural and original disease.... 'Tis by the same vanity of imagination that [man] equals himself to God, attributes to himself divine qualities, and withdraws and separates himself from the crowd of other creatures.[23]

It is surely not a coincidence that the writer who rejects such self-exaltation should also, in his essay "On Cruelty," be among the very few writers since Roman times to assert that cruelty to animals is wrong in itself, quite apart from its tendency to lead to cruelty to human beings.

Perhaps, then, from this point in the development of Western thought the status of nonhumans was bound to improve? The old concept of the universe, and of the central place of human beings in it, was slowly giving ground; modern science was about to set forth on its now-famous rise; and, after all, the status of nonhumans was so low that one might reasonably think it could only improve.

But the absolute nadir was still to come. The last, most bizarre, and—for the animals—most painful outcome of Christian doctrines emerged in the first half of the seventeenth century, in the philosophy of René Descartes. Descartes was a distinctively modern thinker. He is regarded as the father of modern philosophy, and also of analytic geometry, in which a good deal of modern mathematics has its origins. But he was also a Christian, and his beliefs about animals arose from the combination of these two aspects of his thought.

Under the influence of the new and exciting science of mechanics, Descartes held that everything that consisted of matter was governed by mechanistic principles, like those that governed a clock. An obvious problem with this view was our own nature. The human body is composed of matter, and is part of the physical universe. So it would seem that human beings must also be machines, whose behavior is determined by the laws of science.

Descartes was able to escape the unpalatable and heretical view that humans are machines by bringing in the idea of the soul. There are, Descartes said, not one but two kinds of things in the universe, things of the spirit or soul as well as things of a physical or material nature. Human beings are conscious, and consciousness cannot have its origin in matter. Descartes identified consciousness with the immortal soul, which survives the decomposition of the physical body, and asserted that the soul was specially created by God. Of all material beings, Descartes said, only human beings have a soul. (Angels and other immaterial beings have consciousness and nothing else.)

Thus in the philosophy of Descartes the Christian doctrine that animals do not have immortal souls has the extraordinary consequence that they do not have consciousness either. They are, he said, mere machines, automata. They experience neither pleasure nor pain, nor anything else. Although they may squeal when cut with a knife, or writhe in their efforts to escape contact with a hot iron, this does not, Descartes said, mean that they feel pain in these situations. They are governed by the same principles as a clock, and if their actions are more complex than those of a clock, it is because the clock is a machine made by humans, while animals are infinitely more complex machines, made by God.[24]

This "solution" of the problem of locating consciousness in a materialistic world seems paradoxical to us, as it did to many of Descartes's contemporaries, but at the time it was also thought to have important advantages. It provided a reason for believing in a life after death, something which Descartes thought "of great importance" since "the idea that the souls of animals are of the same nature as our own, and that we have no more to fear or to hope for after this life than have the flies and ants" was an error that was apt to lead to immoral conduct. It also eliminated the ancient and vexing theological puzzle of why a just God would allow animals—who neither inherited Adam's sin, nor are recompensed in an afterlife—to suffer.[25]

Descartes was also aware of more practical advantages:

> My opinion is not so much cruel to animals as indulgent to men—at least to those who are not given to the superstitions of Pythagoras—since it absolves them from the suspicion of crime when they eat or kill animals.[26]

For Descartes the scientist the doctrine had still another fortunate result. It was at this time that the practice of experimenting on live animals became widespread in Europe. Since there were no anesthetics then, these experiments must have caused the animals to behave in a way that would indicate, to most of us, that they were suffering extreme pain. Descartes's theory allowed the experimenters to dismiss any qualms they might feel under these circumstances. Descartes himself dissected living animals in order to advance his knowledge of anatomy, and many of the leading physiologists of the period declared themselves Cartesians and mechanists. The following eyewitness account of some of these experimenters, working at the Jansenist seminary of Port-Royal in the late seventeenth century, makes clear the convenience of Descartes's theory:

> They administered beatings to dogs with perfect indifference, and made fun of those who pitied the creatures as if they felt pain. They said the animals were clocks; that the cries they emitted when struck were only the noise of a little spring that had been touched, but that the whole body was without feeling. They nailed poor animals up on boards by

> their four paws to vivisect them and see the circulation of the blood which was a great subject of conversation.[27]

From this point, it really was true that the status of animals could only improve.

The Enlightenment and After

The new vogue for experimenting on animals may itself have been partly responsible for a change in attitudes toward animals, for the experiments revealed a remarkable similarity between the physiology of human beings and other animals. Strictly, this was not inconsistent with what Descartes had said, but it made his views less plausible. Voltaire put it well:

> There are barbarians who seize this dog, who so greatly surpasses man in fidelity and friendship, and nail him down to a table and dissect him alive, to show you the mesaraic veins! You discover in him *all the same organs of feeling as in yourself*. Answer me, mechanist, has Nature arranged all the springs of feeling in this animal *to the end that he might not feel*?[28]

Although no radical change took place, a variety of influences combined to improve attitudes to animals. There was a gradual recognition that other animals do suffer and are entitled to some consideration. It was not thought that they had any rights, and their interests were overridden by human interests; nevertheless the Scottish philosopher David Hume was expressing a common enough sentiment when he said that we are "bound by the laws of humanity to give gentle usage to these creatures."[29]

"Gentle usage" is, indeed, a phrase that nicely sums up the attitude that began to spread in this period: we were entitled to use animals, but we ought to do so gently. The tendency of the age was for greater refinement and civility, more benevolence and less brutality, and animals benefited from this tendency along with humans.

The eighteenth century was also the period in which we rediscovered "Nature": Jean-Jacques Rousseau's noble savage, strolling naked through the woods, picking fruits and nuts as he went, was the culmination of this idealization of nature. By seeing ourselves as part of nature, we regained a sense of kinship with "the beasts." This kinship, however, was in no sense egalitarian. At best, man was seen in the role of benevolent father of the family of animals.

Religious ideas of the special status of human beings did not disappear. They were interwoven with the newer, more benevolent attitude. Alexander Pope, for example, opposed the practice of cutting open fully conscious dogs by arguing that although "the inferior creation" has been "submitted to our power" we are answerable for the "mismanagement" of it.[30]

Finally, and especially in France, the growth of anticlerical feeling was favorable to the status of animals. Voltaire, who delighted in fighting dogmas of all kinds, compared Christian practices unfavorably with those of the Hindu. He went further than the contemporary English advocates of kind treatment when he referred to the barbarous custom of supporting ourselves upon the flesh and blood of beings like ourselves," although apparently he continued to practice this custom himself.[31] Rousseau, too, seems to have recognized the strength of the arguments for vegetarianism without actually adopting the practice; his treatise on education, *Emile,* contains a long and mostly irrelevant passage from Plutarch that attacks the use of animals for food as unnatural, unnecessary, bloody murder.[32]

The Enlightenment did not affect all thinkers equally in their attitudes toward animals. Immanuel Kant, in his lectures on ethics, still told his students:

> So far as animals are concerned, we have no direct duties. Animals are not self-conscious, and are there merely as a means to an end. That end is man.[33]

But in the same year that Kant gave these lectures—1780—Jeremy Bentham completed his *Introduction to the Principles of Morals and Legislation,* and in it, in a passage I have already quoted in the first chapter of this book, he gave the definitive answer to Kant: "The question is not, Can they *reason*? nor Can they *talk*? but, Can they *suffer*?" In comparing the position of animals with that of black slaves, and looking forward to the day "when the rest of the animal creation may acquire those rights which never could have been withholden from them but by the hand of tyranny," Bentham was perhaps the first to denounce "man's dominion" as tyranny rather than legitimate government.

The intellectual progress made in the eighteenth century was followed, in the nineteenth century, by some practical improvements in the conditions of animals. These took the form of laws against wanton cruelty to animals. The first battles for legal rights for animals were fought in Britain, and the initial reaction of the British Parliament indicates that Bentham's ideas had had little impact on his countrymen.

The first proposal for a law to prevent abuse of animals was a bill to prohibit the "sport" of bull-baiting. It was introduced into the House of Commons in 1800. George Canning, the foreign secretary, described it as "absurd" and asked rhetorically: "What could be more innocent than bull-baiting, boxing, or dancing?" Since no attempt was being made to prohibit boxing or dancing, it appears that this astute statesman had missed the point of the bill he was opposing—he thought it an attempt to outlaw gatherings of "the rabble" that might lead to immoral conduct.[34] The presupposition that made this mistake possible was that conduct that injures only an animal cannot possibly be worth legislating about—a presupposition shared by *The Times*, which devoted an editorial to the principle that "whatever meddles with the private personal disposition of man's time or property is tyranny. Till another person is injured there is no room for power to interpose." The bill was defeated.

In 1821 Richard Martin, an Irish gentleman-landowner and a member of Parliament for Galway, proposed a law to prevent the ill-treatment of horses. The following account conveys the tone of the ensuing debate:

> When Alderman C. Smith suggested that protection should be given to asses, there were such howls of laughter that *The Times* reporter could hear little of what was said. When the Chairman repeated this proposal, the laughter was intensified. Another member said Martin would be legislating for dogs next, which caused a further roar of mirth, and a cry "And cats!" sent the House into convulsions.[35]

This bill failed too, but in the following year Martin succeeded with a bill that made it an offense "wantonly" to mistreat certain domestic animals, "the property of any other person or persons." For the first time, cruelty to animals was a punishable offense. Despite the mirth of the previous year, asses were included; dogs and cats, however, were still beyond the pale. More significantly, Martin had had to frame his bill so that it resembled a measure to protect items of private property, for the benefit of the owner, rather than for the sake of the animals themselves.[36]

The bill was now law, but it still had to be enforced. Since the victims could not make a complaint, Martin and a number of other notable humanitarians formed a society to gather evidence and bring prosecutions. So began the first animal welfare organization, later to become the Royal Society for the Prevention of Cruelty to Animals.

A few years after the passage of this first, modest statutory prohibition of cruelty to animals, Charles Darwin wrote in his diary: "Man in his arrogance thinks himself a great work, worthy of the interposition of a deity. More humble and, I believe, true, to consider him created from animals."[37] Another twenty years were to pass before, in 1859, Darwin considered that he had accumulated enough evidence in support of his theory to make it public. Even then, in *The Orgin of Species*, Darwin carefully avoided any discussion of the extent to which his theory of the evolution of one species from another could be applied to humans, saying only that the work would illuminate "the origin of man and his history." In fact, Darwin already had extensive notes on the theory that Homo sapiens had descended from other animals, but he decided that publishing this material would "only add to the prejudices against my views."[38] Only in 1871, when many scientists had accepted the general theory of evolution, did Darwin publish *The Descent of Man*, thus making explicit what had been concealed in a single sentence of his earlier work.

So began a revolution in human understanding of the relationship between ourselves and the nonhuman animals . . . or did it? One would expect the intellectual upheaval sparked by the publication of the theory of evolution to have made a marked difference in human attitudes to animals. Once the weight of scientific evidence in favor of the theory became apparent, practically every earlier justification of our supreme place in creation and our dominion over the animals had to be reconsidered. Intellectually the Darwinian revolution was genuinely revolutionary. Human beings now knew that they were not the special creation of God, made in the divine image and set apart from the animals; on the contrary, human beings came to realize that they were animals themselves. Moreover, in support of his theory of evolution, Darwin pointed out that the differences between human beings and animals were not so great as was generally supposed. Chapter 3 of *The Descent of Man* is devoted to a comparison of the mental powers of humans and the "lower animals," and Darwin summarizes the results of this comparison as follows:

> We have seen that the senses and intuitions, the various emotions and faculties, such as love, memory, attention and curiosity, imitation, reason etc., of which man boasts, may he found in an incipient, or even sometimes in a well-developed condition, in the lower animals.[39]

The fourth chapter of the same work goes still further, affirming that the human moral sense can also be traced back to social instincts in animals that lead them to take pleasure in each other's company, feel sympathy for each other, and perform services of mutual assistance. And in a subsequent work, *The Expression of the Emotions in Man and Animals*, Darwin provided additional evidence of extensive parallels between the emotional life of human beings and that of other animals.

The storm of resistance that met the theory of evolution and of the descent of the human species from animals—a story too well known to need retelling here—is an indication of the extent to which speciesist ideas had come to dominate Western thought. The idea that we are the product of a special act of creation, and that the other animals were created to serve us, was not to be given up without resistance. The scientific evidence for a common origin of the human and other species was, however, overwhelming.

With the eventual acceptance of Darwin's theory we reach a modern understanding of nature, one which has since then changed in detail rather than in fundamentals. Only those who prefer religious faith to beliefs based on reasoning and evidence can still maintain that the human species is the special darling of the entire universe, or that other animals were created to provide us with food, or that we have divine authority over them, and divine permission to kill them.

When we add this intellectual revolution to the growth of humanitarian feeling that preceded it, we might think that all will now be well. Yet, as I hope the preceding chapters have made plain, the human "hand of tyranny" is still clamped down on other species, and we probably inflict more pain on animals now than at any other time in history. What went wrong?

If we look at what relatively advanced thinkers wrote about animals from the time when, toward the end of the eighteenth century, the right of animals to some degree of consideration was beginning to be accepted, we may notice an interesting fact. With very rare exceptions these writers, even the best of them, stop short of the point at which their arguments would lead them to face the choice between breaking the deeply ingrained habit of eating the flesh of other animals or admitting that they do not live up to the conclusions of their own moral arguments. This is an often-repeated pattern. When reading among sources from the late eighteenth century onward, one frequently comes across passages in which the author urges the wrongness of our treatment of other animals in such strong terms that one feels sure that here, at last, is someone who has freed himself altogether from speciesist ideas—and hence, has freed himself too from the most widespread of all speciesist practices, the practice of eating other animals. With one or two notable exceptions (in the nineteenth century Lewis Gompertz and Henry Salt),[40] one is always disappointed. Suddenly a qualification is made, or some new consideration introduced, and the author spares himself the qualms over his diet that his argument seemed sure to create. When the history of the Animal Liberation movement comes to be written, the era that began with Bentham will be known as the era of excuses.

The excuses used vary, and some of them show a certain ingenuity. It is worthwhile examining specimens of the main types, for they are still encountered today.

First, and this should come as no surprise, there is the Divine Excuse. It may be illustrated by the following passage from William Paley's *Principles of Moral and Political Philosophy* (1785). In setting out "the General Rights of Mankind" Paley asks whether we have a right to the flesh of animals:

> Some excuse seems necessary for the pain and loss which we occasion to brutes, by restraining them of their liberty, mutilating their bodies, and at last, putting an end to their lives (which we suppose to be the whole of their existence) for our pleasure or convenience.
>
> [It is] alleged in vindication of this practice . . . that the several species of brutes being created to prey upon one another affords a kind of analogy to prove that the human species were intended to feed upon them . . . [but] the analogy contended for is extremely lame; since brutes have no power to support life by any other means, and since we have; for the whole human species might subsist entirely upon fruits, pulse, herbs and roots, as many tribes of Hindoos actually do. . . .
>
> It seems to me that it would be difficult to defend this right by any arguments which the light and order of nature afford; and that we are beholden for it to the permission recorded in Scripture, *Genesis* ix, 1, 2, 3.[41]

Paley is only one of many who have appealed to revelation when they found themselves unable to give a rational justification of a diet consisting of other animals. Henry Salt in his autobiography *Seventy Years Amongst Savages* (an account of his life in England) records a conversation he had when he was a master at Eton College. He had recently become a vegetarian; now for the first time he was to discuss his practice with a colleague, a distinguished science teacher. With some trepidation he awaited the verdict of the scientific mind on his new beliefs; when it came, it was: "But don't you think that animals were *sent* to us for food?"[42]

Another writer, Lord Chesterfield, appealed to nature, instead of God:

> My scruples remained unreconciled to the committing of so horrid a meal, till upon serious reflection I became convinced of its legality from the general order of nature, which has instituted the universal preying upon the weaker as one of her first principles.[43]

Whether Lord Chesterfield thought this justified cannibalism is not recorded.

Benjamin Franklin used the same argument—the weakness of which Paley exposed—as a justification for returning to a flesh diet after some years as a vegetarian. In his *Autobiography* he recounts how he was watching some friends fishing, and noticed that some of the fish they caught had eaten other fish. He therefore concluded, "If you eat one another, I don't see why we may not eat you." Franklin, however, was at least more honest than some who use this argument, for he admits that he reached this conclusion only after the fish was in the frying-pan and had begun to smell "admirably well"; and he adds that one of the advantages of being a "reasonable creature" is that one can find a reason for whatever one wants to do.[44]

It is also possible for a deep thinker to avoid confronting the troublesome issue of diet by regarding it as altogether too profound for the human mind to comprehend. As Dr. Thomas Arnold of Rugby wrote:

> The whole subject of the brute creation is to me one of such painful mystery that I dare not approach it.[45]

This attitude was shared by the French historian Michelet; being French, he expressed it less prosaically:

> Animal Life, somber mystery! Immense world of thoughts and of dumb sufferings. All nature protests against the barbarity of man, who misapprehends, who humiliates, who tortures his inferior brethren. Life, death! The daily murder which feeding upon animals implies—those hard and bitter problems sternly placed themselves before my mind. Miserable contradiction. Let us hope that there may be another sphere in which the base, the cruel fatalities of this may be spared to us.[46]

Michelet seems to have believed that we cannot live without killing; if so, his anguish at this "miserable contradiction" must have been in inverse proportion to the amount of time he gave to examining it.

Another to accept the comfortable error that we must kill to live was Arthur Schopenhauer. Schopenhauer was influential in introducing Eastern ideas to the West, and in several passages he contrasted the "revoltingly crude" attitudes to animals prevalent in Western philosophy and religion with those of Buddhists and Hindus. His prose is sharp and scornful, and many of his acute criticisms of Western attitudes are still appropriate today. After one particularly biting passage, however, Schopenhauer briefly considers the question of killing for food. He can hardly deny that human beings can live without killing—he knows too much about the Hindus for that—but he claims that "without animal food the human race could not even exist *in the North*." Schopenhauer gives no basis for this geographical distinction, although he does add that the death of the animal should be made "even easier" by means of chloroform.[47]

Even Bentham, who stated so clearly the need to extend rights to nonhumans, flinched at this point:

> There is very good reason why we should be suffered to eat such of them as we like to eat; we are the better for it, and they are never the worse. They have none of those long-protracted anticipations of future misery which we have. The death they suffer in our hands commonly is, and always may be, a speedier, and by that means a less painful one, than that which would await them in the inevitable course of nature.

One cannot help feeling that in these passages Schopenhauer and Bentham lowered their normal standards of argument. Quite apart from the question of the morality of painless killing, neither Schopenhauer nor Bentham considers the suffering necessarily involved in rearing and slaughtering animals on a commercial basis. Whatever the purely theoretical possibilities of painless killing may be, the large-scale killing of animals for food is not and never has been painless. When Schopenhauer and Bentham wrote, slaughter was an even more horrific affair than it is today. The animals were forced to cover long distances on foot, driven to slaughter by drovers who had no concern but to complete the journey as quickly as possible; they might then spend two or three days in the slaughteryards, without food, perhaps without water; they were then slaughtered by barbaric methods, without any form of prior stunning.[48] Despite what Bentham says, they did have some form of anticipation of what was in store for them, at least from the time they entered the slaughteryard and smelled the blood of their fellows. Bentham and Schopenhauer would not, of course, have approved of this, yet they continued to support the process by consuming its products, and justifying the general practice of which it was part. In this respect Paley seems to have had a more accurate conception of what was involved in eating flesh. He, however, could safely look the facts in the face, because he had divine permission to fall back upon; Schopenhauer and Bentham could not have availed themselves of this excuse, and so had to turn their gaze away from the ugly reality.

As for Darwin himself, he too retained the moral attitudes to animals of earlier generations, though he had demolished the intellectual foundations of those attitudes. He continued to dine on the flesh of those beings who, he had said, were capable of love, memory, curiosity, reason, and sympathy for each other; and he refused to sign a petition urging the RSPCA to press for legislative control of experiments on animals.[49] His followers went out of their way to emphasize that although we were a part of nature and descended from animals, our status had not been altered. In reply to the accusation that Darwin's ideas undermined the dignity of man, T. H. Huxley, Darwin's greatest champion, said:

> No-one is more strongly convinced than I am of the vastness of the gulf between civilized man and the brutes; our reverence for the nobility of mankind will not be lessened by the knowledge that man is, in substance and in structure, one with the brutes.[50]

Huxley is a true representative of modern attitudes; he knows perfectly well that the old reasons for assuming a vast gulf between "man" and "brute" no longer stand up, but continues to believe in the existence of such a gulf nevertheless.

Here we see most clearly the ideological nature of our justifications of the use of animals. It is a distinctive characteristic of an ideology that it resists refutation. If the foundations of an ideological position are knocked out from under it, new foundations will be found, or else the ideological position will just hang there, defying the logical equivalent of the laws of gravity. In the case of attitudes to animals, the latter seems to have happened. While the modern view of our place in the world differs enormously from all the earlier views we studied, in the practical matter of how we act toward other animals little has changed. If animals are no longer quite outside the moral sphere, they are still in a special section near the outer rim. Their interests are allowed to count only when they do not clash with human interests. If there is a clash—even a clash between a lifetime of suffering for a nonhuman animal and the gastronomic preference of a human being—the interests of the nonhuman are disregarded. The moral attitudes of the past are too deeply embedded in our thought and our practices to be upset by a mere change in our knowledge of ourselves and of other animals.

Chapter 6

Speciesism Today . . .

defenses, rationalizations, and objections to Animal Liberation and the progress made in overcoming them

We have seen how, in violation of the fundamental moral principle of equality of consideration of interests that ought to govern our relations with all beings, humans inflict suffering on nonhumans for trivial purposes; and we have seen how generation after generation of Western thinkers has sought to defend the right of human beings to do this. In this final chapter I shall look at some of the ways in which speciesist practices are maintained and promoted today, and at the various arguments and excuses that are still used in defense of animal slavery. Some of these defenses have been raised against the position taken in this book, and so this chapter provides an opportunity to answer some of the objections most often made to the case for Animal Liberation; but the chapter is also intended as an extension of the previous one, revealing the continued existence of the ideology whose history we have traced back to the Bible and the ancient Greeks. It is important to expose and criticize this ideology, because although contemporary attitudes to animals are sufficiently benevolent—on a very selective basis—to allow some improvements in the conditions of animals to be made without challenging basic attitudes to animals, these improvements will always be in danger of erosion unless we alter the underlying position that sanctions the ruthless exploitation of nonhumans for human ends. Only by making a radical break with more than two thousand years of Western thought about animals can we build a solid foundation for the abolition of this exploitation.

Our attitudes to animals begin to form when we are very young, and they are dominated by the fact that we begin to eat meat at an early age. Interestingly enough, many children at first refuse to eat animal flesh, and only become accustomed to it after strenuous efforts by their parents, who mistakenly believe that it is necessary for good health. Whatever the child's initial reaction, though, the point to notice is that we eat animal flesh long before we are capable of understanding that what we are eating is the dead body of an animal. Thus we never make a conscious, informed decision, free from the bias that accompanies any long-established habit, reinforced by all the pressures of social conformity, to eat animal flesh. At the same time children have a natural love of animals, and our society encourages them to be affectionate toward animals such as dogs and cats and toward cuddly, stuffed toy animals. These facts help to explain the most distinctive characteristic of the attitudes of children in our society to animals—namely, that rather than having one unified attitude to animals, the child has two conflicting attitudes that coexist, carefully segregated so that the inherent contradiction between them rarely causes trouble.

Not so long ago children were brought up on fairy tales in which animals, especially wolves, were pictured as cunning enemies of man. A characteristic happy ending would leave the wolf drowning in a pond, weighed down by stones which the ingenious hero had sewn into its belly while it was asleep. And in case children missed the implications of these stories, they could all join hands and sing a nursery rhyme like:

Three blind mice. See how they run.
They all ran after the farmer's wife.
She cut off their tails with a carving knife.
Did you ever see such a sight in your life
As three blind mice?

For children brought up on these stories and rhymes there was no inconsistency between what they were taught and what they ate. Today, however, such stories and rhymes have gone out of fashion, and on the surface all is sweetness and light, so far as children's attitudes to animals are concerned. Thereby a problem has arisen: What about the animals we eat?

One response to this problem is simple evasion. The child's affection for animals is directed toward animals that are not eaten: dogs, cats, and other companion animals. These are the animals that an urban or suburban child is most likely to see. Cuddly, stuffed toy animals are more likely to be bears or lions than pigs or cows. When farm animals are mentioned in picture books, stories, and on children's television shows, however, evasion may become a deliberate attempt to mislead children about the nature of modern farms, and so to screen them from the reality that we examined in Chapter 3. An example of this is the popular Hallmark book *Farm Animals*, which presents the child with pictures of hens, turkeys, cows, and pigs, all surrounded by their young, with not a cage, shed, or stall in sight. The text tells us that pigs "enjoy a good meal, then roll in the mud and let out a squeal!" while "Cows don't have a thing to do, but switch their tails, eat grass and moo."[1] British books, like *The Farm* in the best-selling Ladybird series, convey the same impression of rural simplicity, showing the hen running freely in an orchard with her chicks, and all the other animals living with their offspring in spacious quarters.[2] With this kind of early reading it is not surprising that children grow up believing that even if animals "must" die to provide human beings with food, they live happily until that time comes.

Recognizing the importance of the attitudes we form when young, the feminist movement has succeeded in fostering the growth of a new children's literature, in which brave princesses occasionally rescue helpless princes, and girls play the central, active roles that used to be reserved for boys. To alter the stories about animals that we read to our children will not be so easy, since cruelty is not an ideal subject for children's stories. Yet it should be possible to avoid the most gruesome details, and still give children picture books and stories that encourage respect for animals as independent beings, and not as cute little objects that exist for our amusement and table; and as children grow older, they can be made aware that most animals live under conditions that are not very pleasant. The difficulty will be that nonvegetarian parents are going to be reluctant to let their children learn the full story, for fear that the child's affection for animals may disrupt family meals. Even now, one frequently hears that, on learning that animals are killed to provide meat, a friend's child has refused to eat meat. Unfortunately this instinctive rebellion is likely to meet strong resistance from nonvegetarian parents, and most children are unable to keep up their refusal in the face of opposition from parents who provide their meals and tell them that they will not grow up big and strong without meat. One hopes that as knowledge of nutrition spreads more parents will realize that on this issue their children may be wiser than they are.[3] It is an indication of the extent to which people are now isolated from the animals they eat that children brought up on storybooks that lead them to think of a farm as a place where animals wander around freely in idyllic conditions might be able to live out their entire lives without ever being forced to revise this rosy image. There are no farms in the cities and suburbs where people live, and while on a drive through the country one now sees many farm buildings and relatively few animals out in the fields, how many of us can distinguish a storage barn from a broiler shed?

Nor do the mass media educate the public on this topic. American television broadcasts programs on animals in the wild (or supposedly in the wild—sometimes the animals have been captured and released in a more limited space to make filming easier) almost every night of the week; but film of intensive farms is limited to the briefest of glimpses as part of infrequent "specials" on agriculture or food production. The average viewer must know more about the lives of cheetahs and sharks than he or she knows about the lives of chickens or veal calves. The result is that most of the "information" about farm animals to be gained from watching television is in the form of paid advertising, which ranges from ridiculous cartoons of pigs who want to be made into sausages and tuna trying to get themselves canned, to straightforward lies about the conditions in which broiler chickens are reared. The newspapers do little better. Their coverage of nonhuman animals is dominated by "human interest" events like the birth of a baby gorilla at the zoo, or by threats to endangered species; but developments in farming techniques that deprive millions of animals of freedom of movement go unreported.

Before the recent successes of the Animal Liberation movement in exposing one or two notorious laboratories, what went on in research with animals was no better known than what goes on down on the farm. The public, of course, does not have access to laboratories. Although researchers publish their reports in professional journals, researchers only release news of their work to the media when they can claim to have discovered something of special importance. Thus, until the Animal Liberation movement was able to attract national media attention, the public had no idea that most experiments performed on animals are never published at all, and that most of those published are trivial anyway. Since, as we saw in Chapter 2, no one knows exactly how many experiments are performed on animals in the United States, it is not surprising that the public still has not the remotest idea of the extent of animal experimentation. Research facilities are usually designed so that the public sees little of the live animals that go in, or the dead ones that come out. (A standard textbook on the use of animals in experimentation advises laboratories to install an incinerator, since the sight of dozens of bodies of dead animals left out as ordinary refuse "will certainly not enhance the esteem with which the research center or school is held by the public."[4])

Ignorance, then, is the speciesist's first line of defense. Yet it is easily breached by anyone with the time and determination to find out the truth. Ignorance has prevailed so long only because people do not want to find out the truth. "Don't tell me, you'll spoil my dinner" is the usual reply to an attempt to tell someone just how that dinner was produced. Even people who are aware that the traditional family farm has been taken over by big business interests, and that some questionable experiments go on in laboratories, cling to a vague belief that conditions cannot be too bad, or else the government or the animal welfare societies would have done something about it. Some years ago Dr. Bernhard Grzimek, director of the Frankfurt Zoo and one of West Germany's most outspoken opponents of intensive farming, likened the ignorance of Germans about these farms to the ignorance of an earlier generation of Germans to another form of atrocity, also hidden away from most eyes;[5] and in both cases, no doubt, it is not the inability to find out what is going on as much as a desire not to know about facts that may lie heavy on one's conscience that is responsible for the lack of awareness—as well as, of course, the comforting thought that, after all, the victims of whatever it is that goes on in those places are not members of one's own group.

The thought that we can rely on the animal welfare societies to see that animals are not cruelly treated is a reassuring one. Most countries now have at least one large, well-established animal protection society; in the United States there are the American Society for the Prevention of Cruelty to Animals, the American Humane Association, and the Humane Society of the United States; in Britain the Royal Society for the Prevention of Cruelty to Animals remains unchallenged as the largest group. It is reasonable to ask: Why have these associations not been able to prevent the clear cruelties described in Chapters 2 and 3 of this book?

There are several reasons for the failure of the animal welfare establishment to take action against the most important kinds of cruelty. One is historical. When first founded, the RSPCA and ASPCA were radical groups, far ahead of the public opinion of their times, and opposed to all forms of cruelty to animals, including cruelty to farm animals, who then, as now, were the victims of many of the worst abuses. Gradually, however, as these organizations grew in wealth, membership, and respectability, they lost their radical commitment and became part of the "establishment." They built up close contacts with members of the government, and with businessmen and scientists. They tried to use these contacts to improve the conditions of animals, and some minor improvements resulted; but at the same time contacts with those whose basic interests are in the use of animals for food or research purposes blunted the radical criticism of the exploitation of animals that had inspired the founders. Again and again the societies compromised their fundamental principles for the sake of trivial reforms. Better some progress now than nothing at all, they said; but often the reforms proved ineffective in improving the conditions of the animals, and functioned rather to reassure the public that nothing further needed to be done.[6]

As their wealth increased, another consideration became important. The animal welfare societies had been set up as registered charities. This status brought them substantial tax savings; but it is a condition of being registered as a charity, in both Great Britain and the United States, that the charitable organization does not engage in political activities. Political action, unfortunately, is sometimes the only way to improve the conditions of animals (especially if an organization is too cautious to call for public boycotts of animal products), but most of the large groups kept well clear of anything that might endanger their charitable status. This has led them to emphasize safe activities like collecting stray dogs and prosecuting individual acts of wanton cruelty, instead of broad campaigns against systematic cruelty.

Finally, at some point during the last hundred years the major animal welfare societies lost interest in farm animals. Perhaps this was because the supporters and officials of the societies came from the cities and knew more and cared more about dogs and cats than about pigs and calves. Whatever the reason, for most of the present century, the literature and publicity of the old established groups made a significant contribution to the prevailing attitude that dogs and cats and wild animals need protection, but other animals do not. Thus people came to think of "animal welfare" as something for kindly ladies who are dotty about cats, and not as a cause founded on basic principles of justice and morality.

The last decade has seen a change. First, dozens of new, more radical Animal Liberation and animal rights groups have sprung up. Together with some previously existing organizations that had been able to make relatively little impact until then, these new groups have greatly increased public awareness of the immense, systematic cruelty that takes place in intensive animal production, in laboratories, and in circuses, zoos, and hunting. Secondly, perhaps in response to this new wave of interest in the conditions of animals, more established groups such as the RSPCA in Britain and, in America, the ASPCA and Humane Society of the United States, have taken a much more forceful stand against cruelty to farm and laboratory animals, even calling for boycotts of products like intensively produced veal, bacon, and eggs.[7]

Among the factors that make it difficult to arouse public concern about animals perhaps the hardest to overcome is the assumption that "human beings come first" and that any problem about animals cannot be comparable, as a serious moral or political issue, to problems about humans. A number of things can be said about this assumption. First, it is in itself an indication of speciesism. How can anyone who has not made a thorough study of the topic possibly know that the problem is less serious than problems of human suffering? One can claim to know this only if one assumes that animals really do not matter, and that however much they suffer, their suffering is less important than the suffering of humans. But pain is pain, and the importance of preventing unnecessary pain and suffering does not diminish because the being that suffers is not a member of our species. What would we think of someone who said that "whites come first" and that therefore poverty in Africa does not pose as serious a problem as poverty in Europe?

It is true that many problems in the world deserve our time and energy. Famine and poverty, racism, war and the threat of nuclear annihilation, sexism, unemployment, preservation of our fragile environment—all are major issues, and who can say which is the most important? Yet once we put aside speciesist biases we can see that the oppression of nonhumans by humans ranks somewhere along with these issues. The suffering that we inflict on nonhuman beings can be extreme, and the numbers involved are gigantic: more than 100 million pigs, cattle, and sheep go through the processes described in Chapter 3 each year, in the United States alone; billions of chickens do the same; and at least 25 million animals are experimented upon annually. If a thousand human beings were forced to undergo the kind of tests that animals undergo to test the toxicity of household products, there would be a national uproar. The use of millions of animals for this purpose should cause at least as much concern, especially since this suffering is so unnecessary and could easily be stopped if we wanted to stop it. Most reasonable people want to prevent war, racial inequality, poverty, and unemployment; the problem is that we have been trying to prevent these things for years, and now we have to admit that, for the most part, we don't really know how to do it. By comparison, the reduction of the suffering of nonhuman animals at the hands of humans will be relatively easy, once human beings set themselves to do it.

In any case, the idea that "humans come first" is more often used as an excuse for not doing anything about either human or nonhuman animals than as a genuine choice between incompatible alternatives. For the truth is that there is no incompatibility here. Granted, everyone has a limited amount of time and energy, and time taken in active work for one cause reduces the time available for another cause; but there is nothing to stop those who devote their time and energy to human problems from joining the boycott of the products of agribusiness cruelty. It takes no more time to be a vegetarian than to eat animal flesh. In fact, as we saw in Chapter 4, those who claim to care about the well-being of human beings and the preservation of our environment should become vegetarians for that reason alone. They would thereby increase the amount of grain available to feed people elsewhere, reduce pollution, save water and energy, and cease contributing to the clearing of forests; moreover, since a vegetarian diet is cheaper than one based on meat dishes, they would have more money available to devote to famine relief, population control, or whatever social or political cause they thought most urgent. I would not question the sincerity of vegetarians who take little interest in Animal Liberation because they give priority to other causes; but when nonvegetarians say that "human problems come first" I cannot help wondering what exactly it is that they are doing for human beings that compels them to continue to support the wasteful, ruthless exploitation of farm animals.

At this point a historical digression is appropriate. It is often said, as a kind of corollary of the idea that "humans come first," that people in the animal welfare movement care more about animals than they do about human beings. No doubt this is true of some people. Historically, though, the leaders of the animal welfare movement have cared far more about human beings than have other humans who cared nothing for animals. Indeed, the overlap between leaders of movements against the oppression of blacks and women, and leaders of movements against cruelty to animals, is extensive; so extensive as to provide an unexpected form of confirmation of the parallel between racism, sexism, and speciesism. Among the handful of founders of the RSPCA, for example, were William Wilberforce and Fowell Buxton, two of the leaders in the fight against Negro slavery in the British Empire.[8] As for early feminists, Mary Wollstonecraft wrote, in addition to her *Vindication of the Rights of Woman*, a collection of children's stories entitled *Original Stories*, expressly designed to encourage kinder practices toward animals;[9] and a number of the early American feminists, including Lucy Stone, Amelia Bloomer, Susan B. Anthony, and Elizabeth Cady Stanton, were connected with the vegetarian movement. Together with Horace Greeley, the reforming, antislavery editor of *The Tribune*, they would meet to toast "Women's Rights and Vegetarianism."[10]

To the animal welfare movement, too, must go the credit for starting the fight against cruelty to children. In 1874 Henry Bergh, the pioneer of the American animal welfare societies, was asked to do something about a little animal who had been cruelly beaten. The little animal turned out to be a human child; nevertheless Bergh successfully prosecuted the child's custodian for cruelty to an animal, under a New York animal protection statute that he had drafted and bullied the legislature into passing. Further cases were then brought, and the New York Society for the Prevention of Cruelty to Children was set up. When the news reached Britain, the RSPCA set up a British counterpart—the National Society for the Prevention of Cruelty to Children.[11] Lord Shaftesbury was one of the founders of this group. As a leading social reformer, author of the Factory Acts that put an end to child labor and fourteen-hour work days, and a notable campaigner against uncontrolled experimentation and other forms of cruelty to animals, Shaftesbury, like many other humanitarians, clearly refutes the idea that those who care about nonhumans do not care about humans, or that working for one cause makes it impossible to work for the other.

Our conceptions of the nature of nonhuman animals, and faulty reasoning about the implications that follow from our conception of nature, also help to buttress our speciesist attitudes. We have always liked to think ourselves less savage than the other animals. To say that people are "humane" is to say that they are kind; to say that they are "beastly," "brutal," or simply that they behave "like animals" is to suggest that they are cruel and nasty. We rarely stop to consider that the animal who kills with the least reason to do so is the human animal. We think of lions and wolves as savage because they kill; but they must kill, or starve. Humans kill other animals for sport, to satisfy their curiosity, to beautify their bodies, and to please their palates. Human beings also kill members of their own species for greed or power. Moreover, human beings are not content with mere killing. Throughout history they have shown a tendency to torment and torture both their fellow human beings and their fellow animals before putting them to death. No other animal shows much interest in doing this.

While we overlook our own savagery, we exaggerate that of other animals. The notorious wolf for instance, villain of so many folk tales, has been shown by the careful investigations of zoologists in the wild to be a highly social animal, a faithful and affectionate spouse—not just for a season, but for life—a devoted parent, and a loyal member of the pack. Wolves almost never kill anything except to eat it. If males should fight among themselves, the fight ends with a gesture of submission in which the loser offers to his conqueror the underside of his neck—the most vulnerable part of his body. With his fangs only an inch away from the jugular vein of his foe, the victor will be content with submission, and, unlike a human conqueror, does not kill the vanquished opponent.[12]

In keeping with our picture of the world of animals as a bloody scene of combat, we ignore the extent to which other species exhibit a complex social life, recognizing and relating to other members of their species as individuals. When human beings marry, we attribute their closeness to each other to love, and we feel keenly for a human being who has lost his or her spouse. When other animals pair for life, we say that it is just instinct that makes them do so, and if a hunter or trapper kills or captures an animal for research or for a zoo, we do not consider that the animal might have a spouse who will suffer from the sudden absence of the dead or captured animal. In a similar way we know that to part a human mother from her child is tragic for both; but neither farmers nor breeders of companion animals and research animals give any thought to the feelings of the nonhuman mothers and children whom they routinely separate as part of their business.[13]

Curiously, while people often dismiss complex aspects of animal behavior as "mere instinct," and therefore not worthy of comparison with the apparently similar behavior of human beings, these same people will also ignore or overlook the importance of simple instinctive patterns of behavior when it is convenient for them to do so. Thus it is often said of laying hens, veal calves, and dogs kept in cages for experimental purposes that this does not cause them to suffer since they have never known other conditions. We saw in Chapter 3 that this is a fallacy. Animals feel a need to exercise, stretch their limbs or wings, groom themselves, and turn around, whether or not they have ever lived in conditions that permit this. Herd or flock animals are disturbed when they are isolated from others of their species, though they may never have known other conditions, and too large a herd or flock can have the same effect through the inability of the individual animal to recognize other individuals. These stresses reveal themselves in "vices" like cannibalism.

Widespread ignorance of the nature of nonhuman animals allows those who treat animals in this manner to brush off criticism by saying that, after all, "they're not human." Indeed, they are not; but neither are they machines for converting fodder into flesh, or tools for research. Considering how far the knowledge of the general public lags behind the most recent findings of zoologists and ethologists who have spent months and sometimes years observing animals with notebook and camera, the dangers of sentimental anthropomorphism are less serious than the opposite danger of the convenient and self-serving idea that animals are lumps of clay whom we can mold in whatever manner we please.

The nature of nonhuman animals serves as a basis for other attempts to justify our treatment of them. It is often said, as an objection to vegetarianism, that since other animals kill for food, we may do so too. This analogy was already old in 1785, when William Paley refuted it by reference to the fact that while human beings can live without killing, other animals have no choice but to kill if they are to survive.[14] This is certainly true in most cases; a few exceptions may be found—animals who could survive without meat, but eat it occasionally (chimpanzees, for example)—but they are scarcely the species we usually find on our dinner tables. In any case, even if other animals who could live on a vegetarian diet do sometimes kill for food, this would provide no support for the claim that it is morally defensible for us to do the same. It is odd how humans, who normally consider themselves so far above other animals, will, if it seems to support their dietary preferences, use an argument that implies that we ought to look to other animals for moral inspiration and guidance. The point, of course, is that nonhuman animals are not capable of considering the alternatives, or of reflecting morally on the rights and wrongs of killing for food; they just do it. We may regret that this is the way the world is, but it makes no sense to hold nonhuman animals morally responsible or culpable for what they do. Every reader of this book, on the other hand, is capable of making a moral choice on this matter. We cannot evade our responsibility for our choice by imitating the actions of beings who are incapable of making this kind of choice.

(Now, someone is sure to say, I have admitted that there is a significant difference between humans and other animals, and thus I have revealed the flaw in my case for the equality of all animals. Anyone to whom this criticism has occurred should read Chapter 1 more carefully. You will then find that you have misunderstood the nature of the case for equality I made there. I have never made the absurd claim that there are no significant differences between normal adult humans and other animals. My point is not that animals are capable of acting morally, but that the moral principle of equal consideration of interests applies to them as it applies to humans. That it is often right to include within the sphere of equal consideration beings who are not themselves capable of making moral choices is implied by our treatment of young children and other humans who, for one reason or another, do not have the mental capacity to understand the nature of moral choice. As Bentham might have said, the point is not whether they can choose, but whether they can suffer.)

Perhaps the claim is a different one. As we saw in the previous chapter, Lord Chesterfield used the fact that animals eat other animals as a way of arguing that to do so is part of "the general order of nature."[15] He did not indicate why we should imagine that our nature is more like that of the carnivorous tiger than the vegetarian gorilla, or the virtually vegetarian chimpanzee. But quite apart from this objection, we should be wary of appeals to "nature" in ethical argument. Nature may often "know best," but we must use our own judgment in deciding when to follow nature. For all I know, war is "natural" to human beings—it certainly seems to have been a preoccupation for many societies, in very different circumstances, over a long period of history—but I have no intention of going to war to make sure that I act in accordance with nature. We have the capacity to reason about what it is best to do. We should use this capacity (and if you are really keen on appeals to "nature," you can say that it is natural for us to do so).

It must be admitted that the existence of carnivorous animals does pose one problem for the ethics of Animal Liberation, and that is whether we should do anything about it. Assuming that human beings could eliminate carnivorous species from the earth, and that the total amount of suffering among animals in the world would thereby be reduced, should we do it?

The short and simple answer is that once we give up our claim to "dominion" over the other species we should stop interfering with them at all. We should leave them alone as much as we possibly can. Having given up the role of tyrant, we should not try to play God either.

Though it contains part of the truth, this answer is too short and simple. Like it or not, human beings do know more than other animals about what may happen in the future, and this knowledge may put us in a situation in which it would be callous not to interfere. In October 1988, television viewers throughout the world applauded the success of American and Russian efforts to release two California gray whales trapped in the Alaskan ice. Some critics noticed the irony in making such extensive efforts to save two whales, while about two thousand whales are still being killed by human hunters each year, not to mention the estimated 125,000 dolphins who drown annually in the nets of the tuna industry.[16] Nevertheless, it would be a callous person who could assert that the rescue was a bad thing.

So it is conceivable that human interference will improve the conditions of animals, and so be justifiable. But when we consider a scheme like the elimination of carnivorous species, we are considering an entirely different matter. Judging by our past record, any attempt to change ecological systems on a large scale is going to do far more harm than good. For that reason, if for no other, it is true to say that, except in a few very limited cases, we cannot and should not try to police all of nature. We do enough if we eliminate our own unnecessary killing and cruelty toward other animals.[17]

Yet another purported justification of our treatment of animals relies on the fact that in their natural state some animals kill other animals. People often say that, bad as modern farm conditions are, they are no worse than conditions in the wild, where animals are exposed to cold, hunger, and predators; and the implication is that therefore we should not object to modern farm conditions.

Interestingly, defenders of slavery imposed on black Africans often made a similar point. One of them wrote:

> On the whole, since it is evident beyond all controversy that the removal of the Africans, from the state of brutality, wretchedness and misery, in which they are at home so deeply involved, to this land of light, humanity and Christian knowledge, is to them so great a blessing; however faulty any individuals may have been in point of unnecessary cruelty, practised in this business; that, whether the general state of subordination here, which is a necessary consequence of their removal, be agreeable to the law of nature, can by no means longer remain a question.[18]

Now it is difficult to compare two sets of conditions as diverse as those in the wild and those on a factory farm (or those of free Africans and slaves on a plantation); but if the comparison has to be made surely the life of freedom is to be preferred. Factory farm animals cannot walk, run, stretch freely, or be part of a family or herd. True, many wild animals die from adverse conditions or are killed by predators; but animals kept in farms do not live for more than a fraction of their normal life span either. The steady supply of food on a farm is not an unmitigated blessing, since it deprives animals of their most basic natural activity, the search for food. The result is a life of utter boredom, with nothing at all to do but lie in a stall and eat.

In any case, the comparison between factory farm conditions and natural conditions is really irrelevant to the justifiability of factory farms, since this is not the choice that we face. Abolishing factory farms would not mean returning the animals inside them to the wild. Animals in factory farms today were bred by human beings to be raised in these farms and sold for food. If the boycott of factory farm produce advocated in this book is effective, it will reduce the amount of factory farm products that are bought. This does not mean that overnight we will go from the present situation to one in which no one buys these products. (I am optimistic about Animal Liberation, but not totally deluded.) The reduction will be gradual. It will make animal raising less profitable. Farmers will turn to other types of farming, and the giant corporations will invest their capital elsewhere. The result will be that fewer animals will be bred. The number of animals in factory farms will decline because those killed will not be replaced, and not because animals are being sent "back" to the wild. Eventually, perhaps (and now I am allowing my optimism free rein), the only herds of cattle and pigs to be found will be on large reservations, rather like our wildlife refuges. The choice, therefore, is not between life on a factory farm and life in the wild, but whether animals destined to live on factory farms and then killed for food should be born at all.

At this point a further objection may be raised. Noting that if we were all vegetarians there would be far fewer pigs, cattle, chickens, and sheep, a few meat-eaters have claimed that they are actually doing the animals they eat a favor, since but for their desire to eat meat, those animals would never have come into existence at all![19]

In the first edition of this book, I rejected this view on the grounds that it requires us to think that bringing a being into existence confers a benefit on that being—and to hold this, we must believe that it is possible to benefit a nonexistent being. This, I thought, was nonsense. But now I am not so sure. (My unequivocal rejection of this view is, in fact, the only philosophical point made in the earlier edition on which I have changed my mind.) After all, most of us would agree that it would be wrong to bring a child into the world if we knew, before the child was conceived, that it would have a genetic defect that would make its life brief and miserable. To conceive such a child is to cause it harm. So can we really deny that to bring into the world a being who will have a pleasant life is to confer on that being a benefit? To deny this, we would need to explain why the two cases are different, and I cannot find a satisfactory way of doing that.[20]

The argument we are now considering raises the issue of the wrongness of killing—an issue which, because it is so much more complicated than the wrongness of inflicting suffering, I have kept in the background up to this point. Our brief discussion near the close of the first chapter, however, was enough to show that for a being capable of having desires for the future there may be something particularly bad about being killed, something that is not equaled by the creation of another being. The real difficulty arises when we consider beings not capable of having desires for the future—beings who can be thought of as living moment by moment rather than having a continuous mental existence. Granted, even here, killing still seems repugnant. An animal may struggle against a threat to its life, even if it cannot grasp that it has "a life" in the sense that requires an understanding of what it is to exist over a period of time. But in the absence of some form of mental continuity it is not easy to explain why the loss to the animal killed is not, from an impartial point of view, made good by the creation of a new animal who will lead an equally pleasant life.[21]

I still have doubts about this issue. The proposition that the creation of one being should somehow compensate for the death of another does have an air of peculiarity. Of course, if we had a clear basis for saying that all sentient creatures have a right to life (even those not capable of having desires about the future) then it would be easy to say why killing a sentient creature is a kind of wrong that cannot be made good by creating a new creature. But such a position has its own deep philosophical and practical difficulties, as I and others have indicated elsewhere.[22]

On a purely practical level, one can say this: killing animals for food (except when necessary for sheer survival) makes us think of them as objects we can use casually for our own nonessential purposes. Given what we know about human nature, as long as we continue to think of animals in this way we will not succeed in changing the attitudes that, when put into practice by ordinary human beings, lead to disrespect—and hence mistreatment—for the animals. So it might be best to make it a simple general principle to avoid killing animals for food except when it is necessary for survival.

This argument against killing for food relies on a prediction about the consequences of holding an attitude. It is impossible to prove the prediction correct; that is something on which we can only make a judgment on the basis of our knowledge of our fellow human beings. If this prediction is not persuasive, though, the argument we are considering still remains very limited in its application. It certainly does not justify eating meat from factory-produced animals, for they suffer lives of boredom and deprivation, unable to satisfy their basic needs to turn around, groom, stretch, exercise, or take part in the social interactions normal for their species. To bring them into existence for a life of that kind is no benefit to them, but rather a great harm. At the most, the argument from the benefit of bringing a being into existence could justify continuing to eat free-range animals (of a species incapable of having desires for the future), who have a pleasant existence in a social group suited to their behavioral needs, and are then killed quickly and without pain. I can respect conscientious people who take care to eat only meat that comes from such animals—but I suspect that unless they live on a farm where they can look after their own animals, they will, in practice, be very nearly vegetarian anyway.[23]

One final point on the argument that the loss of an animal is compensated for by the creation of a new one. Those who use this ingenious defense of their desire to eat pork or beef rarely follow out its implications. If it were good to bring beings into existence then presumably, other things being equal, we ought to bring as many humans as possible into existence too; and if to this we add the view that human lives are more important than the lives of animals—a view the flesh-eater seems sure to accept—then the argument may be turned on its head, to the discomfort of its original proponent. Since more humans may be fed if we do not feed our grain to livestock, the upshot of the argument is, after all, that we ought to become vegetarians!

Speciesism is so pervasive and widespread an attitude that those who attack one or two of its manifestations—like the slaughter of wild animals by hunters, or cruel experimentation, or bullfighting—often participate in other speciesist practices themselves. This allows those attacked to accuse their opponents of inconsistency. "You say we are cruel because we shoot deer," the hunters say, "but you eat meat. What is the difference, except that you pay someone else to do the killing for you?" "You object to killing animals to clothe ourselves in their skins," say the furriers, "but you are wearing leather shoes." The experimenters plausibly ask why, if people accept the killing of animals to please their palates, they should object to the killing of animals to advance knowledge; and if the objection is just to suffering, they can point out that animals killed for food do not live without suffering either. Even the bullfight enthusiast can argue that the death of the bull in the ring gives pleasure to thousands of spectators, while the death of the steer in a slaughterhouse gives pleasure only to the few people who eat some part of it; and while in the end the bull may suffer more acute pain than the steer, for most of his life it is the bull who is better treated.

The charge of inconsistency really gives no logical support to the defenders of cruel practices. As Brigid Brophy has put it, it remains true that it is cruel to break people's legs, even if the statement is made by someone in the habit of breaking people's arms.[24] Yet people whose conduct is inconsistent with their professed beliefs will find it difficult to persuade others that their beliefs are right; and they will find it even more difficult to persuade others to act on those beliefs. Of course, it is always possible to find some reason for distinguishing between, say, wearing furs and wearing leather: many fur-bearing animals die only after hours or even days spent with a leg caught in a steel-toothed trap, while the animals from whose skins leather is made are spared this agony.[25] There is a tendency, however, for these fine distinctions to blunt the force of the original criticism; and in some cases I do not think distinctions can validly be drawn at all. Why, for instance, is the hunter who shoots a deer for venison subject to more criticism than the person who buys a ham at the supermarket? Overall, it is probably the intensively reared pig who has suffered more.

The first chapter of this book sets out a clear ethical principle—of equal consideration of the interests of all animals—by which we can determine which of our practices affecting nonhuman animals are justifiable and which are not. By applying this principle to our own lives we can make our actions fully consistent. Thus we can deny to those who ignore the interests of animals the opportunity to charge us with inconsistency.

For all practical purposes as far as urban and suburban inhabitants of the industrialized nations are concerned, following the principle of equal consideration of interests requires us to be vegetarians. This is the most important step, and the one to which I have given most attention; but we should also, to be consistent, stop using other animal products for which animals have been killed or made to suffer. We should not wear furs. We should not buy leather products either, since the sale of hides for leather plays a significant role in the profitability of the meat industry.

For the pioneer vegetarians of the nineteenth century, giving up leather meant a real sacrifice, since shoes and boots made of other materials were scarce. Lewis Gompertz, the second secretary of the RSPCA and a strict vegetarian who refused to ride in horse-drawn vehicles, suggested that animals should be reared in pastures and allowed to grow old and die a natural death, after which their skins would be used for leather.[26] The idea is a tribute to Gompertz's humanity rather than his economics, but today the economics are on the other foot. Shoes and boots made of synthetic materials are now available in many cheaper stores, at prices considerably lower than those for leather shoes; and sneakers made of canvas and rubber are now the standard footwear for American youth. Belts, bags, and other goods once made of leather are now easily found in other materials.

Other problems that used to daunt the most advanced opponents of the exploitation of animals have also disappeared. Candles, once made only of tallow, are no longer indispensable, and can, for those who still want them, be obtained in nonanimal materials. Soaps made from vegetable oils rather than animal fats are obtainable from health food stores. We can do without wool, and although sheep generally roam freely, there is a strong case for doing so in view of the many cruelties to which these gentle animals are subjected.[27] Cosmetics and perfumes, often made from wild animals like the musk deer and the Ethiopian civet cat, are hardly essential items anyway, but those who wish to wear them can obtain cruelty-free cosmetics, which do not contain animal products and have not been tested on animals either, from a number of shops and organizations.[28]

Although I mention these alternatives to animal products to show that it is not difficult to refuse to participate in the major kinds of exploitation of animals, I do not believe that consistency is the same as, or implies, a rigid insistence on standards of absolute purity in all that one consumes or wears. The point of altering one's buying habits is not to keep oneself untouched by evil, but to reduce the economic support for the exploitation of animals, and to persuade others to do the same. So it is not a sin to continue to wear leather shoes you bought before you began to think about Animal Liberation. When your leather shoes wear out, buy nonleather ones; but you will not reduce the profitability of killing animals by throwing out your present ones. With diet, too, it is more important to remember the major aims than to worry about such details as whether the cake you are offered at a party was made with a factory farm egg.

We are still a long way from the point at which it is possible to put pressure on restaurants and food manufacturers to eliminate animal products altogether. That point will come when a significant section of the population is boycotting meat and other factory farm products. Until then consistency demands only that we do not contribute significantly to the demand for animal products. Thus we can demonstrate that we have no need of animal products. We are more likely to persuade others to share our attitude if we temper our ideals with common sense than if we strive for the kind of purity that is more appropriate to a religious dietary law than to an ethical and political movement.

Usually it is not too difficult to be consistent in one's attitudes to animals. We do not have to sacrifice anything essential, because in our normal life there is no serious clash of interests between human and nonhuman animals. It must be admitted, though, that it is possible to think of more unusual cases in which there is a genuine clash of interests. For instance, we need to grow crops of vegetables and grain to feed ourselves; but these crops may be threatened by rabbits, mice, or other "pests." Here we have a clear conflict of interest between humans and nonhumans. What would be done about it, if we were to act in accordance with the principle of equal consideration of interests?

First let us note what is done about this situation now. The farmer will seek to kill off the "pests" by the cheapest method available. This is likely to be poison. The animals will eat poisoned baits, and die a slow, painful death. No consideration at all is given to the interests of the "pests"—the very word "pest" seems to exclude any concern for the animals themselves.[29] But the classification "pest" is our own, and a rabbit that is a pest is as capable of suffering, and as deserving of consideration, as a white rabbit who is a beloved companion animal. The problem is how to defend our own essential food supplies while respecting the interests of these animals to the greatest extent possible. It should not be beyond our technological abilities to find a solution to this problem which, if not totally satisfactory to all concerned, at least causes far less suffering than the present "solution." The use of baits that cause sterility, instead of a lingering death, would be an obvious improvement.

When we have to defend our food supplies against rabbits, or our houses and our health against mice and rats, it is as natural for us to lash out violently at the animals that invade our property as it is for the animals themselves to seek food where they can find it. At the present stage of our attitudes to animals, it would be absurd to expect people to change their conduct in this respect. Perhaps in time, however, when more major abuses have been remedied, and attitudes to animals have changed, people will come to see that even animals who are in some sense "threatening" our welfare do not deserve the cruel deaths we inflict upon them; and so we may eventually develop more humane methods of limiting the numbers of those animals whose interests are genuinely incompatible with our own.

A similar reply may be given to those hunters and controllers of what are misleadingly called "wildlife refuges" who claim that to prevent overpopulation by deer, seals, or whatever the animal in question may be, hunters must periodically be allowed to "harvest" the excess population—this allegedly being in the interests of the animals themselves. The use of the term "harvest"—often found in the publications of the hunters' organizations—gives the lie to the claim that this slaughter is motivated by concern for the animals. The term indicates that the hunter thinks of deer or seals as if they were corn or coal, objects of value only in so far as they serve human interests. This attitude, which is shared to a large extent by the U.S. Fish and Wildlife Service, overlooks the vital fact that deer and other hunted animals are capable of feeling pleasure and pain. They are therefore not means to our ends, but beings with interests of their own. If it is true that in special circumstances their population grows to such an extent that they damage their own environment and the prospects of their own survival, or that of other animals who share their habitat, then it may be right for humans to take some supervisory action; but obviously if we consider the interests of the animals, this action will not be to allow hunters to kill some animals, inevitably wounding others in the process, but rather to reduce the fertility of the animals. If we made an effort to develop more humane methods of population control for wild animals in reserves, it would not be difficult to come up with something better than what is done now. The trouble is that the authorities responsible for wildlife have a "harvest" mentality, and are not interested in finding techniques of population control that would reduce the number of animals to be "harvested" by hunters.[30]

I have said that the difference between animals like deer—or pigs and chickens, for that matter—whom we ought not to think of "harvesting," and crops like corn, which we may harvest, is that the animals are capable of feeling pleasure and pain, while the plants are not. At this point someone is bound to ask: "How do we know that plants do not suffer?"

This objection may arise from a genuine concern for plants; but more often those raising it do not seriously contemplate extending consideration to plants if it should be shown that they suffer; instead they hope to show that if we were to act on the principle I have advocated we would have to stop eating plants as well as animals, and so would starve to death. The conclusion they draw is that if it is impossible to live without violating the principle of equal consideration, we need not bother about it at all, but may go on as we have always done, eating plants and animals.

The objection is weak in both fact and logic. There is no reliable evidence that plants are capable of feeling pleasure or pain. Some years ago a popular book, *The Secret Life of Plants*, claimed that plants have all sorts of remarkable abilities, including the ability to read people's minds. The most striking experiments cited in the book were not carried out at serious research institutions, and attempts by researchers in major universities to repeat the experiments have failed to obtain any positive results. The book's claims have now been completely discredited.[31]

In the first chapter of this book I gave three distinct grounds for believing that nonhuman animals can feel pain: behavior, the nature of their nervous systems, and the evolutionary usefulness of pain. None of these gives us any reason to believe that plants feel pain. In the absence of scientifically credible experimental findings, there is no observable behavior that suggests pain; nothing resembling a central nervous system has been found in plants; and it is difficult to imagine why species that are incapable of moving away from a source of pain or using the perception of pain to avoid death in any other way should have evolved the capacity to feel pain. Therefore the belief that plants feel pain appears to be quite unjustified.

So much for the factual basis of this objection. Now let us consider its logic. Assume that, improbable as it seems, researchers do turn up evidence suggesting that plants feel pain. It would still not follow that we may as well eat what we have always eaten. If we must inflict pain or starve, we would then have to choose the lesser evil. Presumably it would still be true that plants suffer less than animals, and therefore it would still be better to eat plants than to eat animals. Indeed this conclusion would follow even if plants were as sensitive as animals, since the inefficiency of meat production means that those who eat meat are responsible for the indirect destruction of at least ten times as many plants as are vegetarians! At this point, I admit, the argument becomes farcical, and I have pursued it this far only to show that those who raise this objection but fail to follow out its implications are really just looking for an excuse to go on eating meat.

Up to this point we have been examining, in this chapter, attitudes that are shared by many people in Western societies, and the strategies and arguments that are commonly used to defend these attitudes. We have seen that from a logical point of view these strategies and arguments are very weak. They are rationalizations and excuses rather than arguments. It might be thought, however, that their weakness is due to some lack of expert knowledge that ordinary people have in discussing ethical questions. For that reason, in the first edition of this book I examined what some of the leading philosophers of the 1960s and early 1970s had said about the moral status of nonhuman animals. The results were not to the credit of philosophy.

Philosophy ought to question the basic assumptions of the age. Thinking through, critically and carefully, what most of us take for granted is, I believe, the chief task of philosophy, and the task that makes philosophy a worthwhile activity. Regrettably, philosophy does not always live up to its historic role. Aristotle's defense of slavery will always stand as a reminder that philosophers are human beings and are subject to all the preconceptions of the society to which they belong. Sometimes they succeed in breaking free of the prevailing ideology; more often they become its most sophisticated defenders.

So it was with the philosophers of the period just before the first edition of this book appeared. They did not challenge anyone's preconceptions about our relations with other species. By their writings, most philosophers who tackled problems that touched upon the issue revealed that they made the same unquestioned assumptions as most other human beings, and what they said tended to confirm readers in their comfortable speciesist habits.

At that time, discussions of equality and rights in moral and political philosophy were almost always formulated as problems of human equality and human rights. The effect of this was that the issue of the equality of animals never confronted philosophers or their students as an issue in itself—already an indication of the failure of philosophy up to that time to probe accepted beliefs. Yet philosophers found it difficult to discuss the issue of human equality without raising questions about the status of nonhumans. The reason for this—which may already be apparent from the first chapter of this book—has to do with the way in which the principle of equality must be interpreted and defended, if it is to be defended at all.

For philosophers of the 1950s and 1960s, the problem was to interpret the idea that all human beings are equal in a manner that does not make it plainly false. In most ways, human beings are not equal; and if we seek some characteristic that all of them possess, then this characteristic must be a kind of lowest common denominator, pitched so low that no human being lacks it. The catch is that any such characteristic that is possessed by all human beings will not be possessed only by human beings. For example, all human beings, but not only human beings, are capable of feeling pain; and while only human beings are capable of solving complex mathematical problems, not all humans can do this. So it turns out that in the only sense in which we can truly say, as an assertion of fact, that all humans are equal, at least some members of other species are also "equal"—equal, that is, to some humans.

If, on the other hand, we decide that, as I argued in Chapter 1, these characteristics are really irrelevant to the problem of equality, and equality must be based on the moral principle of equal consideration of interests rather than on the possession of some characteristic, it is even more difficult to find some basis for excluding animals from the sphere of equality.

This result is not what the egalitarian philosophers of that period originally intended to assert. Instead of accepting the outcome to which their own reasonings naturally pointed, however, they tried to reconcile their beliefs in human equality and animal inequality by arguments that are either devious or myopic. For instance, one philosopher prominent in philosophical discussions of equality at the time was Richard Wasserstrom, then professor of philosophy and law at the University of California, Los Angeles. In his article "Rights, Human Rights and Racial Discrimination," Wasserstrom defined "human rights" as those rights that human beings have and nonhumans do not have. He then argued that there are human rights to well-being and to freedom. In defending the idea of a human right to well-being, Wasserstrom said that to deny someone relief from acute physical pain makes it impossible for that person to live a full or satisfying life. He then went on: "In a real sense, the enjoyment of these goods differentiates human from nonhuman entities."[32] The problem is that when we look back to find to what the expression "these goods" refers, the only example given is relief from acute physical pain—something that nonhumans may appreciate as well as humans. So if human beings have a right to relief from acute physical pain, it would not be a specifically human right, in the sense Wasserstrom had defined. Animals would have it too.

Faced with a situation in which they saw a need for some basis for the moral gulf that is still commonly thought to separate human beings and animals, but unable to find any concrete difference between human beings and animals that would do this without undermining the equality of human beings, philosophers tended to waffle. They resorted to high-sounding phrases like "the intrinsic dignity of the human individual."[33] They talked of "the intrinsic worth of all men" (sexism was as little questioned as speciesism) as if all men (humans?) had some unspecified worth that other beings do not have.[34] Or they would say that human beings, and only human beings, are "ends in themselves" while "everything other than a person can only have value for a person."[35]

As we saw in the preceding chapter, the idea of a distinctive human dignity and worth has a long history. In the present century, until the 1970s, philosophers had cast off the original metaphysical and religious shackles of this idea, and freely invoked it without feeling any need to justify the idea at all. Why should we not attribute "intrinsic dignity" or "intrinsic worth" to ourselves? Why should we not say that we are the only things in the universe that have intrinsic value? Our fellow human beings are unlikely to reject the accolades we so generously bestow upon them, and those to whom we deny the honor are unable to object. Indeed, when we think only of human beings it can be very liberal, very progressive, to talk of the dignity of all of them. In so doing we implicitly condemn slavery, racism, and other violations of human rights. We admit that we ourselves are in some fundamental sense on a par with the poorest, most ignorant members of our own species. It is only when we think of human beings as no more than a small subgroup of all the beings that inhabit our planet that we may realize that in elevating our own species we are at the same time lowering the relative status of all other species.

The truth is that the appeal to the intrinsic dignity of human beings appears to solve the egalitarian philosopher's problems only as long as it goes unchallenged. Once we ask why it should be that all human beings—including infants, the intellectually disabled, criminal psychopaths, Hitler, Stalin, and the rest—have some kind of dignity or worth that no elephant, pig, or chimpanzee can ever achieve, we see that this question is as difficult to answer as our original request for some relevant fact that justifies the inequality of humans and other animals. In fact, these two questions are really one: talk of intrinsic dignity or moral worth does not help, because any satisfactory defense of the claim that all and only human beings have intrinsic dignity would need to refer to some relevant capacities or characteristics that only human beings have, in virtue of which they have this unique dignity or worth. To introduce ideas of dignity and worth as a substitute for other reasons for distinguishing humans and animals is not good enough. Fine phrases are the last resource of those who have run out of arguments.

In case anyone still thinks it may be possible to find some relevant characteristic that distinguishes all human beings from all members of other species, let us consider again the fact that there are some human beings who quite clearly are below the level of awareness, self-consciousness, intelligence, and sentience of many nonhuman beings. I am thinking of human beings with severe and irreparable brain damage, and also of infant human beings; to avoid the complication of the potential of infants, however, I shall concentrate on permanently and profoundly retarded human beings.

Philosophers who set out to find a characteristic that would distinguish human beings from other animals rarely took the course of abandoning these groups of human beings by lumping them in with other animals. It is easy to see why they did not do so; to take this line without rethinking our attitudes to other animals would mean we have the right to perform painful experiments on retarded humans for trivial reasons; similarly it would follow that we have the right to rear and kill them for food.

For philosophers discussing the problem of equality, the easiest way out of the difficulty posed by the existence of human beings who are profoundly and permanently disabled intellectually was to ignore it. The Harvard philosopher John Rawls, in his long book *A Theory of Justice*, came up against this problem when trying to explain why we owe justice to human beings but not to other animals, but he brushed it aside with the remark, "I cannot examine this problem here, but I assume that the account of equality would not be materially affected."[36] This is an extraordinary way of handling the issue of equal treatment: it would appear to imply either that we may treat people who are profoundly and permanently disabled intellectually as we now treat animals, or that, contrary to Rawls's own statements, we do owe justice to animals.

What else could philosophers do? If they honestly confronted the problem posed by the existence of human beings with no morally relevant characteristics not also possessed by nonhuman beings, it would be impossible to cling to the equality of human beings without suggesting a radical revision in the status of nonhumans. In a desperate attempt to save the usually accepted views, it was even argued that we should treat beings according to what is "normal for the species" rather than according to their actual characteristics.[37] To see how outrageous this is, imagine that at some future date evidence were to be found that, even in the absence of any cultural conditioning, it was normal for more females than males in a society to stay at home looking after the children instead of going out to work. This finding would, of course, be perfectly compatible with the obvious fact that there are some women who are less well suited to looking after children, and better suited to going out to work, than some men. Would any philosopher then claim that these exceptional women should be treated in accordance with what is "normal for the sex"—and therefore, say, not be admitted to medical school—rather than in accordance with their actual characteristics? I do not think so. I find it hard to see anything in this argument except a defense of preferring the interests of members of our own species because they are members of our own species.

Like the other philosophical arguments common before the idea of equality for animals was taken seriously by philosophers, this one stands as a warning of the ease with which not only ordinary people, but also those most skilled in moral reasoning, can fall victim to a prevailing ideology. Now, however, I am truly delighted to report that philosophy has thrown off its ideological blinkers. Many of today's university courses in ethics really do challenge their students to rethink their attitudes on a range of ethical issues, and the moral status of nonhuman animals is prominent among them. Fifteen years ago I had to search hard to find a handful of references by academic philosophers on the issue of the status of animals; today I could have filled this entire book with an account of what has been written on this topic during the past fifteen years. Articles on how we ought to treat animals are included in virtually all the standard collections of readings used in applied ethics courses. It is the complacent, unargued assumptions of the moral insignificance of nonhuman animals which have become scarce.

In fact over the last fifteen years, academic philosophy has played a major role in fostering and supporting the Animal Liberation movement. The amount of activity can be seen by a glance at Charles Magel's recent bibliography of books and articles on animal rights and related issues. From ancient times to the beginning of the 1970s, Magel finds only 95 works worthy of mention, and of these only two or three are by professional philosophers. During the next eighteen years, however, Magel finds 240 works on animals rights, many by philosophers teaching in universities.[38] Moreover, published works are only part of the story; in philosophy departments all over the United States, Australia, Britain, Canada, and in many other countries too, philosophers are teaching their students about the moral status of animals. Many of them are also working actively for change with animal rights groups, either on campus or off.

Of course, philosophers are not unanimous in support of vegetarianism and Animal Liberation—when were they ever unanimous about anything? But even those philosophers who have been critical of claims made by their colleagues on behalf of animals have accepted important elements of the case for change. For example R. G. Frey of Bowling Green State University, Ohio, who has written more in opposition to my views on animals than any other philosopher, begins one of his articles by stating flatly: "I am not an antivivisectionist..." But he then acknowledges that:

> I have and know of nothing which enables me to say, a priori, that a human life of any quality, however low, is more valuable than an animal life of any quality, however high.

As a result, Frey recognizes that "the case for antivivisectionism is far stronger than most people allow." He concludes that if one seeks to justify experimenting on nonhuman animals by the benefits they produce (which is, in his view, the only way in which the practice can be justified), there is no intrinsic reason why such benefits would not also justify experiments on "humans whose quality of life is exceeded by or equal to that of animals." Hence he accepts experiments on animals where the benefits are sufficiently important, but only at the price of accepting the possibility of similar experiments on humans.[39]

More dramatic still was the change of heart shown by the Canadian philosopher Michael Allen Fox. In 1986 the publication of his book *The Case for Animal Experimentation* seemed sure to earn him a prominent spot at scholarly conferences as the chief philosophical defender of the animal research industry. The drug companies and lobbyists for animal experiments who thought they had, at last, a tame philosopher they could use to defend themselves against ethical criticism must have been dismayed however, when Fox suddenly disavowed his own book. In a response to a highly critical review in *The Scientist*, Fox wrote a letter to the editor saying that he agreed with the reviewer: he had come to see that the arguments of his book were mistaken, and it was not possible to justify animal experimentation on ethical grounds. Later Fox followed through on his courageous change of mind by becoming a vegetarian.[40]

The rise of the Animal Liberation movement may be unique among modern social causes in the extent to which it has been linked with the development of the issue as a topic of discussion in the circles of academic philosophy. In considering the status of nonhuman animals, philosophy itself has undergone a remarkable transformation: it has abandoned the comfortable conformism of accepted dogma and returned to its ancient Socratic role.

The core of this book is the claim that to discriminate against beings solely on account of their species is a form of prejudice, immoral and indefensible in the same way that discrimination on the basis of race is immoral and indefensible. I have not been content to put forward this claim as a bare assertion, or as a statement of my own personal view, which others may or may not choose to accept. I have argued for it, appealing to reason rather than to emotion or sentiment. I have chosen this path, not because I am unaware of the importance of kind feelings and sentiments of respect toward other creatures, but because reason is more universal and more compelling in its appeal. Greatly as I admire those who have eliminated speciesism from their lives purely because their sympathetic concern for others reaches out to all sentient creatures, I do not think that an appeal to sympathy and good-heartedness alone will convince most people of the wrongness of speciesism. Even where other human beings are concerned, people are surprisingly adept at limiting their sympathies to those of their own nation or race. Almost everyone, however, is at least nominally prepared to listen to reason. Admittedly, there are some who flirt with an excessive subjectivism in morality, saying that any morality is as good as any other; but when these same people are pressed to say if they think the morality of Hitler, or of the slave traders, is as good as the morality of Albert Schweitzer or Martin Luther King, they find that, after all, they believe some moralities are better than others.

So throughout this book I have relied on rational argument. Unless you can refute the central argument of this book, you should now recognize that speciesism is wrong, and this means that, if you take morality seriously, you should try to eliminate speciesist practices from your own life, and oppose them elsewhere. Otherwise no basis remains from which you can, without hypocrisy, criticize racism or sexism.

I have generally avoided arguing that we ought to be kind to animals because cruelty to animals leads to cruelty to human beings. Perhaps it is true that kindness to human beings and to other animals often go together; but whether or not this is true, to say, as Aquinas and Kant did, that this is the real reason why we ought to be kind to animals is a thoroughly speciesist position. We ought to consider the interests of animals because they have interests and it is unjustifiable to exclude them from the sphere of moral concern; to make this consideration depend on beneficial consequences for human beings is to accept the implication that the interests of animals do not warrant consideration for their own sakes.

Similarly, I have avoided an extensive discussion of whether a vegetarian diet is healthier than a diet that includes animal flesh. A good deal of evidence suggests that it is, but I have contented myself with showing that a vegetarian can expect to be at least as healthy as one who eats meat. Once one goes beyond this it is difficult to avoid giving the impression that if further studies should show that a diet containing flesh is acceptable from the point of view of health, then the case for becoming a vegetarian collapses. From the standpoint of Animal Liberation, however, so long as we can live without inflicting miserable lives on animals, that is what we ought to do.

I believe that the case for Animal Liberation is logically cogent, and cannot be refuted; but the task of overthrowing speciesism in practice is a formidable one. We have seen that speciesism has historical roots that go deep into the consciousness of Western society. We have seen that the elimination of speciesist practices would threaten the vested interests of the giant agribusiness corporations, and the professional associations of research workers and veterinarians. When necessary, these corporations and organizations are prepared to spend millions of dollars in defense of their interests, and the public will then be bombarded with advertisements denying allegations of cruelty. Moreover the public has—or thinks it has—an interest in the continuance of the speciesist practice of raising and killing animals for food and this makes people ready to accept reassurances that, in this respect at least, there is little cruelty. As we have seen, people are also ready to accept fallacious forms of reasoning, of the type we have examined in this chapter, which they would never entertain for a moment were it not for the fact that these fallacies appear to justify their preferred diet.

Against these ancient prejudices, powerful vested interests, and ingrained habits, does the Animal Liberation movement have any chance at all? Other than reason and morality, does it have anything in its favor? A decade ago there was no concrete basis for hope that its arguments could prevail, other than confidence in the ultimate victory of reason and morality. Since then the movement has seen dramatic growth in the number of its supporters, its public visibility, and most important, the list of gains made for animals. Ten years ago the Animal Liberation movement was widely seen as crackpot, and the membership of groups with a genuinely liberationist philosophy was tiny. Today People for the Ethical Treatment of Animals has 250,000 members, and the Humane Farming Association, which is campaigning strongly against veal crates, has 45,000.[41] Trans-Species Unlimited has grown from a tiny group with one office in central Pennsylvania to a national organization with branches in New York City, New Jersey, Philadelphia, and Chicago. The Coalition to Abolish the LD50 and Draize Tests brings together animal rights and animal welfare groups with a combined membership in the millions. In 1988 Animal Liberation achieved what has become a badge of recognition: a respectful cover story in *Newsweek*.[42]

We have noted some of the gains for animals as they arose in our discussion of particular topics, but they are worth bringing together. They include the prohibition of veal crates in Britain and the phasing out of battery cages in Switzerland and the Netherlands, as well as the more far-reaching legislation in Sweden which will eliminate veal crates, battery cages, sow stalls, and all other devices that prevent animals from moving about freely. It will also make it illegal to keep cattle without allowing them to graze in pastures during the warmer months. The worldwide campaign against the fur trade has succeeded in greatly reducing the quantity of fur sold, especially in Europe. In Britain the House of Fraser, a leading chain of department stores, was the target of protests against fur. In December 1989, it announced that it was closing the fur salons in fifty-nine of its sixty stores, leaving only one remaining in the famous London store, Harrods.

In the United States there have been no gains for farm animals yet, but several particularly objectionable series of experiments have been brought to a halt. The first success was achieved in 1977, when a campaign led by Henry Spira persuaded the American Museum of Natural History to stop a pointless series of experiments that involved mutilating cats in order to investigate the effect this had on their sex lives.[43] In 1981 came exposure by the Animal Liberation activist Alex Pacheco of the appalling conditions of seventeen monkeys at Edward Taub's Institute for Behavioral Research, in Silver Springs, Maryland. The National Institutes of Health cut off Taub's funding, and Taub became the first in the United States to be convicted of cruelty—although the conviction was later reversed on the technical ground that animal experimenters receiving federal tax funding do not have to obey state anticruelty laws.[44] Meanwhile the case gave national prominence to a fledgling group called People for the Ethical Treatment of Animals, which in 1984 led efforts to stop Dr. Thomas Gennarelli's head injury experiments on monkeys at the University of Pennsylvania. These efforts were triggered by extraordinary videotapes of animal abuse, shot by the experimenters themselves and stolen from the laboratory in a night raid carried out by the Animal Liberation Front. Gennarelli's grant was withdrawn.[45] In 1988, after months of picketing by Trans-Species Unlimited, a researcher at Cornell University gave up a $530,000 grant to study barbiturate addiction using cats.[46] Around the same time Benetton, the Italian fashion chain, announced that it would no longer perform safety tests for new cosmetics and toiletries on animals. Benetton had been the target of an international campaign, coordinated by People for the Ethical Treatment of Animals, that involved Animal Liberationists in seven countries. Noxell Corporation, an American cosmetics manufacturer, had not been the object of such a campaign; but it made its own decision to rely on tissue culture instead of performing Draize tests on rabbits to determine whether its products can damage the human eye. Noxell's decision was part of a steady movement toward alternatives by major cosmetics, toiletries, and pharmaceutical corporations, initiated and constantly spurred on by the Coalition to Abolish the LD50 and Draize Tests.[47] Years of hard work paid off in 1989 when Avon, Revlon, Fabergé, Mary Kay, Amway, Elizabeth Arden, Max Factor, Christian Dior, and several small companies announced that they were ending, or at least suspending, all animal experimentation. In the same year the European Commission, which is responsible for safety testing in ten nations of the European Community, announced that it would accept alternatives to the LD50 and Draize tests, and invited all OECD nations (a group which includes the U.S. and Japan) to work toward developing common alternative safety tests. Both the LD50 and the Draize test have now been banned by government regulation in Victoria and New South Wales, Australia's most populous states and the states in which most animal experimentation has been carried out.[48]

In the United States, momentum is also building on the issue of dissection in high schools. The stubborn resistance to dissection of one Californian high school student, Jenifer Graham—and her insistence on not losing marks for her conscientious objection—led to the passage, in 1988, of the California Students' Rights Bill, which gives students in Californian primary and secondary schools the right to refuse to dissect without suffering a penalty. Similar bills are now being introduced in New Jersey, Massachusetts, Maine, Hawaii, and several other states.

As the movement gains increased visibility and support, the groundswell of people doing their part gathers momentum. Rock musicians have helped to circulate the Animal Liberation message. Movie stars, models, and dress designers have pledged to avoid furs. The international success of the Body Shop chain has made cruelty-free cosmetics more attractive and readily available. Vegetarian restaurants are proliferating and even nonvegetarian restaurants are offering vegetarian dishes. All this makes it easier for newcomers to join those already doing what they can to limit cruelty to animals in their daily lives.

Nevertheless, Animal Liberation will require greater altruism on the part of human beings than any other liberation movement. The animals themselves are incapable of demanding their own liberation, or of protesting against their condition with votes, demonstrations, or boycotts. Human beings have the power to continue to oppress other species forever, or until we make this planet unsuitable for living beings. Will our tyranny continue, proving that morality counts for nothing when it clashes with self-interest, as the most cynical of poets and philosophers have always said? Or will we rise to the challenge and prove our capacity for genuine altruism by ending our ruthless exploitation of the species in our power, not because we are forced to do so by rebels or terrorists, but because we recognize that our position is morally indefensible?

The way in which we answer this question depends on the way in which each one of us, individually, answers it.

Appendix 1

Further Reading

This is not a complete list of sources—references to which will be found in the footnotes—but a select list of especially valuable books, including books on the arguments for vegetarianism. Books on vegetarian cooking are listed in Appendix 2. For a detailed, annotated guide to the literature, see Charles Magel's magnificently comprehensive *Keyguide to Information Sources in Animal Rights* (London: Mansell, and Jefferson, North Carolina: McFarland, 1989).

General

Godlovitch, Stanley and Roslind, and John Harris, eds., *Animals, Men and Morals*. New York, Grove, 1974. A path-breaking collection of articles.

Gompertz, Lewis. *Moral Inquiries on the Situation of Man and of Brutes*. London, 1824. One of the first carefully argued proposals for a radically different attitude to animals.

Gruen, Lori, Peter Singer, and David Hine. *Animal Liberation: A Graphic Guide*. London, Camden Press, 1987. An illustrated brief popular account of the Animal Liberation movement in theory and practice.

Midgley, Mary. *Animals and Why They Matter*. Athens, University of Georgia Press, 1984. A penetrating discussion of the difference that species makes.

Rachels, James. *Created from Animals: The Moral Implications of Darwinism*. Oxford and New York, Oxford University Press, 1990. Expounds the still largely unrecognized moral implications of the theory of evolution in regard to our treatment of animals.

Regan, Tom. *The Case For Animal Rights*. Berkeley, University of California Press, 1983. The fullest elaboration of the philosophical arguments for attributing rights to animals.

Regan, Tom and Peter Singer, eds. *Animal Rights and Human Obligations*. Englewood Cliffs, N.J., Prentice-Hall, 2nd ed., 1989. An anthology of writings, both old and new, from both sides.

Rollin, Bernard. *The Unheeded Cry*, Oxford, Oxford University Press, 1989. A highly readable account of attempts to deny feelings to animals and why they fail.

Ryder, Richard D. *Animal Revolution: Changing Attitudes Towards Speciesism*. Oxford, Blackwell, 1989. A historical survey of changing attitudes toward animals, emphasizing the past two decades, by a leading thinker and activist of this period.

Salt, Henry. *Animal's Rights Considered in Relation to Social Progress*. Clarks Summit, Pennsylvania, Society for Animal Rights; Fontwell, Sussex, Centaur Press/State Mutual Book, 1985 (first published 1892). An early classic, now available once more.

Sapontzis, Steve. *Morals, Reason and Animals*. Philadelphia, Temple University Press, 1987. A detailed philosophical analysis of arguments about Animal Liberation.

Singer, Peter, ed. *In Defense of Animals*. New York, Harper and Row, 1986. Specially written essays by leading activists and thinkers.

Thomas, Keith, *Man and the Natural World: Changing Attitudes in England 1500–1800*. London, Allen Lane, 1983. A scholarly but highly readable study of attitudes to animals during this period.

Turner, E. S. *All Heaven in a Rage*. London, Michael Joseph, 1964. An informative and entertaining history of the movement for animal protection.

Wynne-Tyson, J., ed. *The Extended Circle: A Commonplace Book of Animal Rights*. New York, Paragon House, 1988; London, Penguin, 1989. Hundreds of short extracts from humane thinkers down the ages.

Animals in Research

Rowan, Andrew. *Of Mice, Models, and Men: A Critical Evaluation of Animal Research*. Albany, State University of New York Press, 1984. An up-to-date examination by a scientist.

Ryder, Richard. *Victims of Science*. Fontwell, Sussex, Centaur Press/State Mutual Book, 1983. Still among the best overall accounts of animal experimentation.

Sharpe, Robert. *The Cruel Deception*. Wellingborough, Northants, Thorsons, 1988. The scientific case against animal experimentation, arguing that it is generally irrelevant to advances in health, and sometimes positively misleading.

Sperlinger, David, ed. *Animals in Research: New Perspectives in Animal Experimentation*. Chichester and New York, John Wiley, 1983. A collection of essays by scientists and philosophers.

Farm Animals and the Meat Industry

Agriculture Committee, House of Commons. *Animal Welfare in Poultry, Pig and Veal Calf Production*. London, Her Majesty's Stationery Office, 1981. An authoritative government report that comes out firmly against many current practices.

Brambell, F. W. R., Chairman. *Report of the Technical Committee to Enquire into the Welfare of Animals Kept Under Intensive Livestock Husbandry Systems*. London, Her Majesty's Stationery Office, 1965. The report of the first detailed government inquiry into factory farming.

Dawkins, Marian. *Animal Suffering: The Science of Animal Welfare*. New York, Routledge, Chapman and Hall, 1980. A scientific discussion of ways of objectively measuring animal suffering.

Druce, Clare. *Chicken and Egg: Who Pays the Price?* London, Merlin, 1989. An indictment of the egg and poultry industries, with answers to the excuses made on their behalf.

Fox, Michael W. *Farm Animals: Husbandry, Behavior, and Veterinary Practice*. Baltimore, University Park Press, 1984. A documented account of the welfare problems of farm animals.

Gold, Mark. *Assault and Battery*. London, Pluto Press, 1983. A critical examination of factory farming.

Harrison, Ruth. *Animal Machines*. London, Vincent Stuart, 1964. The book that started the campaign against factory farming.

Mason, Jim and Peter Singer. *Animal Factories*. New York, Crown, 1980. The health, ecological, and animal welfare implications of factory farming, with an outstanding collection of photographs.

Schell, Orville. *Modern Meat*. New York, Random House, 1984. A leading journalist looks at "the pharmaceutical farm" and its products.

Vegetarianism

Akers, Keith. *A Vegetarian Sourcebook: The Nutrition, Ecology, and Ethics of a Natural Foods Diet*. Arlington, Va., Vegetarian Press, 1989. A comprehensive collection of up-to-date scientific information on the vegetarian diet.

Gold, Mark. *Living Without Cruelty*. Basingstoke, Hants, Green Print, 1988. Covers all the detailed questions of living without abusing animals.

Kapleau, Roshi P. *To Cherish All Life: A Buddhist View of Animal Slaughter and Meat Eating*. Rochester, N.Y., The Zen Center, 1981. By an eminent American Buddhist.

Lappé, Francis Moore. *Diet for a Small Planet*. New York, Ballantine, 10th Anniversary ed., 1985. This book argues on ecological grounds against meat production.

Moran, Victoria. *Compassion the Ultimate Ethic: An Exploration of Veganism*. Wellingborough, Northants, Thorsons, 1985. How to live without exploiting animals, with discussion of ecological and health aspects of veganism as well.

Robbins, John. *Diet for a New America: How Your Food Choices Affect Your Health, Happiness and the Future of Life on Earth*. Walpole, N.H., Stillpoint, 1987. The author assembles a mass of evidence against eating animal products.

Wynne-Tyson, Jon. *Food For a Future: How World Hunger Could Be Ended by the Twenty-first Century*. Wellingborough, Northants, Thorsons, rev. ed., 1988. An argument for vegetarianism on humane and ecological grounds.

Wildlife

Amory, Cleveland. *Man Kind?* New York, Dell, 1980. A scathing critique of the war on wildlife.

Batten, Peter. *Living Trophies*. New York, Crowell, 1976. A critical look at zoos and what they do to animals.

Day, David. *The Doomsday Book of Animals*. New York, Viking Press, 1980. How we are driving numerous species to extinction.

McKenna, Virginia, Will Travers, and Jonathan Wray, eds. *Beyond the Bars*. Wellingborough, Northants, Thorsons, 1988. A collection of essays on zoos and related issues, focusing especially on wildlife conservation.

Regenstein, Lewis. *The Politics of Extinction*. New York, Macmillan, 1975. An account of how we have driven, and continue to drive, species to extinction.

Appendix 2

Living Without Cruelty

Cruelty-free Products

It is now relatively easy to obtain cruelty-free soaps, toiletries and cosmetics. The worldwide Body Shop chain stocks a wide range of products neither tested on animals nor using animal ingredients. Several other brands are available from health food stores. Cruelty-free products are also advertised in the pages of magazines like *The Animals' Agenda*. Many of the larger organizations will supply a list of approved cruelty-free products; in particular, write to: Beauty Without Cruelty (see addresses under Australia, United Kingdom, and United States in the list below) or in the United Kingdom to the British Union for the Abolition of Vivisection, or the Vegan Society, and in the United States to the National Anti-Vivisection Society, or to People for the Ethical Treatment of Animals. See also Mark Gold's *Living Without Cruelty*, listed in the section of the reading list on vegetarianism.

Those interested in vegetarian diets for dogs or cats should contact Harbingers of a New Age, Box 146, Swisshome, OR 97480, USA, but veterinary advice should also be obtained.

Food

When the first edition of this book appeared there was so little information generally available on vegetarianism that it was necessary to provide a detailed appendix containing nutritional information and vegetarian recipes. Fortunately excellent vegetarian cookbooks are now available in most bookshops, and the nutritional adequacy of a vegetarian diet is not in dispute. So I shall simply list a few of the better cookbooks. For further information on nutritional aspects of living without cruelty, see the books listed above under vegetarianism, especially Keith Akers' *A Vegetarian Sourcebook.*

Bloodroot Collective. *The Political Palate: A Feminist Vegetarian Cookbook.* Bridgeport, Conn., Sanguinaria Publishing, 1980.

————. *The Second Seasonal Political Palate.* Bridgeport, Conn., Sanguinaria Publishing, 1984.

Dinshah, Freya. *The Vegan Cookbook.* American Vegan Society; address on page 265. Just what the name implies.

Ewald, Ellen Buchman. *Recipes for a Small Planet.* New York, Ballantine, 1988. A sequel to Francis Moore Lappé's *Diet for a Small Planet;* more about protein, and a lot more recipes.

Grossinger, Jennie. *The Art of Jewish Cooking.* New York, Random House, 1958. Not vegetarian, but you will find enough vegetarian recipes to keep you gaining weight happily, if you like Jewish food.

Hagler, Louise, ed. *Tofu Cookery.* Summertown, Tenn., The Book Publishing Co., 1982. If you are not yet using tofu regularly, buy this book. You will be amazed at what can be done with this versatile food.

Hurd, Dr. Frank and Mrs. *Ten Talents Vegetarian Natural Foods Cookbook.* Published by the authors, Chisholm, Minn.; available through the Seventh Day Adventist Church. A natural food cookbook and health manual. It is more expensive than the other books on this list, but worthwhile for vegans, since most of the recipes in it are vegan.

Jaffrey, Madhur. *Madhur Jaffrey's World-of-the-East Vegetarian Cooking.* New York, Knopf, 1984.

Katzen, Mollie. *Moosewood Cookbook.* Berkeley, Calif., Ten Speed Press, 1977. Recipes from a celebrated vegetarian restaurant.

————. *The Enchanted Broccoli Forest.* Berkeley, Calif., Ten Speed Press, 1982. The sequel to the *Moosewood Cookbook.*

Lager, Mildred and Dorothea van Gundy Jones. *The Soybean Cookbook.* New York, Arco, 1968; also published in paperback by Arc Books. Three hundred and fifty recipes for using soybeans, including directions for growing bean sprouts and making bean curd.

Lappé, Frances Moore and Ellen Buchman Ewald. *Great Meatless Meals.* New York, Ballantine, 1981. Thirty complete menus, with recipes, for excellent and nutritious vegetarian meals.

Lemlin, Jeanne. *Vegetarian Pleasures: A New Cookbook.* New York, Knopf, 1986. Another cookbook arranged by menus, full of tempting culinary inventions.

Roden, Claudia. *A Book of Middle Eastern Food.* New York, Knopf, 1974; also published in paperback by Vintage Books. Not vegetarian, but has recipes for hummus, felafel, and stuffed vegetables of all kinds. Especially worthwhile for those who use a ground beef substitute, since many of the meat dishes use ground or minced meat, and work well with textured vegetable protein instead.

Thomas, Anna. *The Vegetarian Epicure.* New York, Knopf 1972; also published in paperback by Vintage Books. Many delicious gourmet recipes. Especially good on breads and soups. Not so good for "vegans," as there is too much emphasis on cheese and egg recipes.

Appendix 3

Organizations

Living a cruelty-free lifestyle is important, but it is equally essential to work actively for an end to animal exploitation. Some people can do this by working alone, or forming their own group with a few like-minded friends; but another way is to join an existing organization. I have included here a few active and effective organizations, and some magazines (in italics) promoting an Animal Liberation viewpoint. Organizations come and go, and the fact that an organization is not listed here should not be taken to mean that it is not useful.

Australia

Animal Liberation
State offices are as follows:

Canberra, PO Box 1875, ACT 2601
New South Wales, 20 Enmore Rd., Newtown, NSW 2402
Northern Territory, PO Box 49277, Casuarina, NT 5792
Queensland, GPO Box 1787, Brisbane, Qld. 4001
South Australia, PO Box 114, Rundle Mall, Adelaide, SA 5000
Tasmania, 102 Bathurst St., Hobart, Tas. 7000
Victoria, GPO Box 1196 K, Melbourne, Vic. 3001
Western Australia, PO Box 146, Inglewood, WA 6052

Animal Liberation: The Magazine
PO Box 221
Mitcham, Vic. 3132

Australian and New Zealand Federation of Animal Societies
PO Box 1023
Collingwood, Vic. 3066

Beauty Without Cruelty
GPO Box 1787
Brisbane, Qld. 4001

Canada

ARK II—Canadian Animal Rights Network
542 Mt. Pleasant Road #104
Toronto, Ontario M4S 2M7

Lifeforce
PO Box 3117
Main Post Office
Vancouver, BC V6B 3X6

Germany

Mobilisation für Tiere e.V.
Postfach 977
3400 Göttingen

Verein Gegen Tierquälerische Massentierhaltung e.V.
Teichtor 10
2305 Heikendorf b. Kiel

Israel

Animal Liberation
PO Box 519
Givatayim 53104

Italy

Etica & Animali
Via Marradi, 2
20123 Milano

Mexico

Associacion de Lucha para Evitar la Crueldad con los Animales
Av. Presidente Masarik 350-201
Col. Polanco Chapoltepec
11560 Mexico SDF

Netherlands

Nederlandse Bond tot Bestrijding van de Vivisectie
Jan van Nassaustraat 81
2596 BR's-Gravenhage

New Zealand

Save Animals from Exploitation
PO Box 30139
Takapuna North
Auckland 9

Sweden

Nordiska Samfundet Mot Plagsamma Djurforsok
Drottninggatan 102
11160 Stockholm

Switzerland

Konsumenten Arbeitsgruppe zur Förderung tierfreundlicher umweltgerechter Nutzung von Haustieren (KAG)
Engelgasse 12a
9001 St. Gallen

Stiftung Fonds für versuchstierfreie Forschung FFVFF
Biberlingstr. 5
8032 Zurich

United Kingdom

Animal Aid
7 Castle Street
Tonbridge
Kent TN91BH
Publishes *Outrage*

Animal Liberation Front
BCM Box 1160
London WC1N 3XX

Beauty Without Cruelty
11 Lime Hill Road
Tunbridge Wells
Kent TN1 ILJ

Beauty Without Cruelty
King Henry's Walk
London N14 NH

British Union for the Abolition of Vivisection
16a Crane Grove, Islington
London N7 8LB
Publishes *Liberator*

Chicken's Lib
PO Box 2, Holmfirth
Huddersfield HD7 1QT

Compassion in World Farming
20 Lavant St.
Petersfield
Hampshire GU32 3EW
Publishes *Agscene*

Vegan Society
33-35 George St.
Oxford OX1 2AY

Vegetarian Society
Parkdale, Dunham Road
Altrincham
Cheshire WA14 4QG

United States

American Vegan Society
Box H
Malaga, NJ 08328

The Animals' Agenda
456 Monroe Turnpike
Monroe, CT 06468

The Animals' Voice
PO Box 341347
Los Angeles, CA 90034

The Animal Legal Defense Fund
1363 Lincoln Avenue
San Rafael, CA 94901

Animal Rights Coalitions
(Coalition to Abolish the LD50 and Draize Tests,
and Coalition for Nonviolent Food)
Box 214 Planetarium Station
New York, NY 10024

Association of Veterinarians for Animal Rights
15 Dutch St., Suite 500-A
New York, NY 10038-3779

Beauty Without Cruelty
175 West 12th St., #16G
New York, NY 10011

Between the Species
PO Box 254
Berkeley, CA 94701

CEASE
(Coalition to End Animal Suffering and Exploitation)
PO Box 27
Cambridge, MA 02238

Farm Animal Reform Movement
PO Box 70123
Washington, DC 20088

The Fund for Animals
200 W. 57th Street
New York, NY 10019

Humane Farming Association
1550 California St.
San Francisco, CA 94109

The International Primate Protection League
PO Box 766
Summerville, SC 29484

International Society for Animal Rights
421 South State St.
Clarks Summit, PA 18411

National Anti-vivisection Society
53 West Jackson Blvd., Suite 1550
Chicago, IL 60604

People for the Ethical Treatment of Animals
PO Box 42516
Washington, DC 20015

Physicians Committee for Responsible Medicine
PO Box 6322
Washington, DC 20015

Psychologists for the Ethical Treatment of Animals
PO Box 87
New Gloucester, ME 04260

Trans-Species Unlimited
PO Box 1553
Williamsport, PA 17703

Trans-Species Unlimited
New York Office
PO Box 20697
Columbus Circle Station
New York, NY 10023

United Action for Animals
205 East 42nd St.
New York, NY 10017

Vegetarian Times
PO Box 570
Oak Park, IL 60603

Notes

Chapter 1

1. For Bentham's moral philosophy, see his *Introduction to the Principles of Morals and Legislation,* and for Sidgwick's see *The Methods of Ethics,* 1907 (the passage is quoted from the seventh edition; reprint, London: Macmillan, 1963), p. 382. As examples of leading contemporary moral philosophers who incorporate a requirement of equal consideration of interests, see R.M. Hare, *Freedom and Reason* (New York: Oxford University Press, 1963), and John Rawls, *A Theory of Justice* (Cambridge: Harvard University Press, Belknap Press, 1972). For a brief account of the essential agreement on this issue between these and other positions, see R. M. Hare, "Rules of War and Moral Reasoning," *Philosophy and Public Affairs* 1 (2) (1972).

2. Letter to Henry Gregoire, February 25, 1809.

3. Reminiscences by Francis D. Gage, from Susan B. Anthony, *The History of Woman Suffrage,* vol. 1; the passage is to be found in the extract in Leslie Tanner, ed., *Voices From Women's Liberation* (New York: Signet, 1970).

4. I owe the term "speciesism" to Richard Ryder. It has become accepted in general use since the first edition of this book, and now appears in *The Oxford English Dictionary,* second edition (Oxford: Clarendon Press, 1989).

5. *Introduction to the Principles of Morals and Legislation,* chapter 17.

6. See M. Levin, "Animal Rights Evaluated," *Humanist* 37: 14–15 (July/August 1977); M. A. Fox, "Animal Liberation: A Critique," *Ethics* 88: 134–138 (1978); C. Perry and G. E. Jones, "On Animal Rights," *International Journal of Applied Philosophy* 1: 39–57 (1982).

7. Lord Brain, "Presidential Address," in C.A. Keele and R. Smith, eds., *The Assessment of Pain in Men and Animals* (London: Universities Federation for Animal Welfare, 1962).

8. Lord Brain, "Presidential Address," p. 11.

9. Richard Serjeant, *The Spectrum of Pain* (London: Hart Davis, 1969), p. 72.

10. See the reports of the Committee on Cruelty to Wild Animals (Command Paper 8266, 1951), paragraphs 36–42; the Departmental Committee on Experiments on Animals (Command Paper 2641, 1965), paragraphs 179–182; and the Technical Committee to Enquire into the Welfare of Animals Kept under Intensive Livestock Husbandry Systems (Command Paper 2836, 1965), paragraphs 26–28 (London: Her Majesty's Stationery Office).

11. See Stephen Walker, *Animal Thoughts* (London: Routledge and Kegan Paul, 1983); Donald Griffin, *Animal Thinking* (Cambridge: Harvard University Press, 1984); and Marian Stamp Dawkins, *Animal Suffering: The Science of Animal Welfare* (London: Chapman and Hall, 1980).

12. See Eugene Linden, *Apes, Men and Language* (New York: Penguin, 1976); for popular accounts of some more recent work, see Erik Eckholm, "Pygmy Chimp Readily Learns Language Skill," *The New York Times*, June 24, 1985; and "The Wisdom of Animals," *Newsweek*, May 23, 1988.

13. *In the Shadow of Man* (Boston: Houghton Mifflin, 1971), p. 225. Michael Peters makes a similar point in "Nature and Culture," in Stanley and Roslind Godlovitch and John Harris, eds., *Animals, Men and Morals* (New York: Taplinger, 1972). For examples of some of the inconsistencies in denials that creatures without language can feel pain, see Bernard Rollin, *The Unheeded Cry: Animal Consciousness, Animal Pain, and Science* (Oxford: Oxford University Press, 1989).

14. I am here putting aside religious views, for example the doctrine that all and only human beings have immortal souls, or are

made in the image of God. Historically these have been very important, and no doubt are partly responsible for the idea that human life has a special sanctity. (For further historical discussion see Chapter 5.) Logically, however, these religious views are unsatisfactory, since they do not offer a reasoned explanation of why it should be that all humans and no nonhumans have immortal souls. This belief too, therefore, comes under suspicion as a form of speciesism. In any case, defenders of the "sanctity of life" view are generally reluctant to base their position on purely religious doctrines, since these doctrines are no longer as widely accepted as they once were.

15. For a general discussion of these questions, see my *Practical Ethics* (Cambridge: Cambridge University Press, 1979), and for a more detailed discussion of the treatment of handicapped infants, see Helga Kuhse and Peter Singer, *Should the Baby Live*? (Oxford: Oxford University Press, 1985).

16. For a development of this theme, see my essay, "Life's Uncertain Voyage," in P. Pettit, R. Sylvan and J. Norman, eds., *Metaphysics and Morality* (Oxford: Blackwell, 1987), pp. 154–172.

17. The preceding discussion, which has been changed only slightly since the first edition, has often been overlooked by critics of the Animal Liberation movement. It is a common tactic to seek to ridicule the Animal Liberation position by maintaining that, as an animal experimenter put it recently, "Some of these people believe that every insect, every mouse, has as much right to life as a human." (Dr. Irving Weissman, as quoted in Katherine Bishop, "From Shop to Lab to Farm, Animal Rights Battle is Felt," *The New York Times,* January 14, 1989.) It would be interesting to see Dr. Weissman name some prominent Animal Liberationists who hold this view. Certainly (assuming only that he was referring to the right to life of a human being with mental capacities very different from those of the insect and the mouse) the position described is not mine. I doubt that it is held by many—if any—in the Animal Liberation movement.

Chapter 2

1. U.S. Air Force, School of Aerospace Medicine, Report No. USAFSAM-TR-82-24, August 1982.

2. U.S. Air Force, School of Aerospace Medicine, Report No. USAFSAM-TR-87-19, October 1987.

3. U.S. Air Force, Report No. USAFSAM-TR-87-19, p. 6

4. Donald J. Barnes, "A Matter of Change," in Peter Singer, ed., *In Defense of Animals* (Oxford: Blackwell, 1985).

5. *Air Force Times*, November 28, 1973; *The New York Times*, November 14, 1973.

6. B. Levine et al., "Determination of the Chronic Mammalian Toxicological Effects of TNT: Twenty-six Week Subchronic Oral Toxicity Study of Trinitrotoluene (TNT) in the Beagle Dog," Phase II, Final Report (U.S. Army Medical Research and Development Command, Fort Detrick, Maryland, June 1983).

7. Carol G. Franz, "Effects of Mixed Neutron-gamma Total-body Irradiation on Physical Activity Performance of Rhesus Monkeys," *Radiation Research* 101: 434–441 (1985).

8. *Proceedings of the National Academy of Science* 54: 90 (1965).

9. *Engineering and Science* 33: 8 (1970).

10. *Maternal Care and Mental Health*, World Health Organization Monograph Series, 2: 46 (1951).

11. *Engineering and Science* 33: 8 (1970).

12. *Journal of Comparative and Physiological Psychology* 80 (1): 11 (1972).

13. *Behavior Research Methods and Instrumentation* 1: 247 (1969).

14. *Journal of Autism and Childhood Schizophrenia* 3 (3): 299 (1973).

15. *Journal of Comparative Psychology* 98: 35-44 (1984).

16. *Developmental Psychology* 17: 313–318 (1981).

17. *Primates* 25: 78–88 (1984).

18. Research figures compiled by Martin Stephens, Ph.D., as reported in *Maternal Deprivation Experiments in Psychology: A Critique of Animal Models*, a report prepared for the American, National and New England Anti-Vivisection Societies (Boston, 1986).

19. *Statistics of Scientific Procedures on Living Animals, Great Britain, 1988,* Command Paper 743 (London: Her Majesty's Stationery Office, 1989).

20. U.S. Congress Office of Technology Assessment, *Alternatives to Animal Use in Research, Testing and Education* (Washington, D.C.: Government Printing Office, 1986), p. 64.

21. Hearings before the Subcommittee on Livestock and Feed Grains of the Committee on Agriculture, U.S. House of Representatives, 1966, p. 63.

22. See A. Rowan, *Of Mice, Models and Men* (Albany: State Univeristy of New York Press, 1984), p. 71; his later revision is in a personal communication to the Office of Technology Assessment; see *Alternatives to Animal Use in Research, Testing and Education*, p. 56.

23. OTA, *Alternatives to Animal Use in Research, Testing and Education*, p. 56.

24. *Experimental Animals* 37: 105 (1988).

25. *Nature* 334: 445 (August 4, 1988).

26. *The Harvard Bioscience Whole Rat Catalog* (South Natick, Mass.: Harvard Bioscience, 1983).

27. Report of the Littlewood Committee, pp. 53, 166; quoted by Richard Ryder, "Experiments on Animals," in Stanley and Roslind Godlovitch and John Harris, eds., *Animals, Men, and Morals* (New York: Taplinger, 1972), p. 43.

28. Figures calculated by Lori Gruen from data reports provided by the U.S. Public Health Service, *Computer Retrieval of Information on Scientific Projects* (CRISP) reports.

29. *Journal of Comparative and Physiological Psychology* 67 (1): 110 (April 1969).

30. *Bulletin of the Psychonomic Society* 24: 69–71 (1986).

31. *Behavioral and Neural Biology* 101: 296–299 (1987).

32. *Pharmacology, Biochemistry, and Behavior* 17: 645–649 (1982).

33. *Journal of Experimental Psychology: Animal Behavior Processes* 10: 307–323 (1984).

34. *Journal of Abnormal and Social Psychology* 48 (2): 291 (April 1953).

35. *Journal of Abnormal Psychology* 73 (3): 256 (June 1968).

36. *Animal Learning and Behavior* 12: 332–338 (1984).

37. *Journal of Experimental Psychology: Animal Behavior and Processes* 12: 277–290 (1986).

38. *Psychological Reports* 57: 1027–1030 (1985).

39. *Progress in Neuro-Psychopharmacology and Biological Psychiatry* 8: 434–446 (1984).

40. *Journal of the Experimental Analysis of Behavior* 19 (1): 25 (1973).

41. *Journal of the Experimental Analysis of Behavior* 41: 45–52 (1984).

42. *Aggressive Behavior* 8: 371–383 (1982).

43. *Animal Learning and Behavior* 14: 305–314 (1986).

44. *Behavioral Neuroscience* 100 (2): 90–99 and 98 (3): 541–555 (1984).

45. OTA, *Alternatives to Animal Use in Research, Testing and Education*, p. 132.

46. A. Heim, *Intelligence and Personality* (Baltimore: Penguin, 1971) p. 150; for a splendid discussion of the entire phenomenon, see Bernard Rollin, *The Unheeded Cry: Animal Consciousness, Animal Pain, and Science* (New York: Oxford University Press, 1989).

47. Chris Evans, "Psychology Is About People," *New Scientist*, August 31, 1972, p. 453.

48. *Statistics of Scientific Procedures on Living Animals, Great Britain, 1988* (London: Her Majesty's Stationery Office, 1989), tables 7, 8, and 9.

49. J. P. Griffin and G. E. Diggle, *British Journal of Clinical Pharmacology* 12: 453–463 (1981).

50. OTA, *Alternatives to Animal Use in Research, Testing and Education*, p. 168.

51. *Journal of the Society of Cosmetic Chemists* 13: 9 (1962).

52. OTA, *Alternatives to Animal Use in Research, Testing and Education*, p. 64.

53. *Toxicology* 15 (1): 31–41 (1979).

54. David Bunner et al., "Clinical and Hematologic Effects of T-2 Toxin on Rats," Interim Report (U.S. Army Medical Research and Development Command, Fort Detrick, Frederick, Maryland, Au-

gust 2, 1985). The quotations from the Department of State are from *Report to the Congress for Secretary of State Alexander Haig, March 22, 1982: Chemical Warfare in S.E. Asia and Afghanistan* (U.S. Department of State Special Report No. 98, Washington, D.C., 1982).

55. M. N. Gleason et al., eds., *Clinical Toxicology of Commercial Products* (Baltimore: Williams and Wilkins, 1969).

56. *PCRM Update* (Newsletter of the Physicians Committee for Responsible Medicine, Washington, D.C.), July–August 1988, p. 4.

57. S. F. Paget, ed., *Methods in Toxicology* (Blackwell Scientific Publications, 1970), pp. 4, 134–139.

58. *New Scientist*, March 17, 1983.

59. On Practolol, see W. H. Inman and F. H. Goss, eds., *Drug Monitoring* (New York: Academic Press, 1977); on Zipeprol, see C. Moroni et al., *The Lancet*, January 7, 1984, p. 45. I owe these references to Robert Sharpe, *The Cruel Deception* (Wellingborough, Northants: Thorsons, 1988).

60. S. F. Paget, ed., *Methods in Toxicology*, p. 132.

61. G. F. Somers, *Quantitative Method in Human Pharmacology and Therapeutics* (Elmsford, N.Y.: Pergamon Press, 1959), quoted by Richard Ryder, *Victims of Science* (Fontwell, Sussex: Centaur Press/State Mutual Book, 1983), p. 153.

62. Syndicated article appearing in *West County Times* (California), January 17, 1988.

63. As reported in *DVM: The Newsmagazine of Veterinary Medicine* 9: 58 (June 1988).

64. *The New York Times*, April 15, 1980.

65. For further details see Henry Spira, "Fighting to Win," in Peter Singer, ed., *In Defense of Animals*.

66. *PETA News* (People for the Ethical Treatment of Animals, Washington, D.C.) 4 (2): 19 (March/April 1989).

67. "Noxell Significantly Reduces Animal Testing," News Release, Noxell Corporation, Hunt Valley, Maryland, December 28, 1988; Douglas McGill, "Cosmetics Companies Quietly Ending Animal Tests," *The New York Times*, August 2, 1989, p. 1.

68. "Avon Validates Draize Substitute," News Release, Avon Products, New York, April 5, 1989.

69. *The Alternatives Report* (Center for Animals and Public Policy, Tufts School of Veterinary Medicine, Grafton, Massachusetts) 2: 2 (July/August 1989); "Facts about Amway and the Environment," Amway Corporation, Ada, Michigan, May 17, 1989.

70. "Avon Announces Permanent End to Animal Testing," News Release, Avon Products, New York, June 22, 1989.

71. Douglas McGill, "Cosmetics Companies Quietly Ending Animal Tests," *The New York Times*, August 2, 1989, p. 1.

72. "Industry Toxicologists Keen on Reducing Animal Use," *Science*, April 17, 1987.

73. Barnaby J. Feder, "Beyond White Rats and Rabbits," *The New York Times*, February 28, 1988, Business section, p. 1; see also Constance Holden, "Much Work But Slow Going on Alternatives to Draize Test," *Science*, October 14, 1985, p. 185.

74. Judith Hampson, "Brussels Drops Need for Lethal Animal Tests," *New Scientist*, October 7, 1989.

75. Coalition to Abolish LD50, Coordinators Report 1983 (New York, 1983), p. 1.

76. H. C. Wood, *Fever: A Study of Morbid and Normal Physiology*, Smithsonian Contributions to Knowledge, No. 357 (Lippincott, 1880).

77. *The Lancet*, September 17, 1881, p. 551.

78. *Journal of the American Medical Association* 89 (3): 177 (1927).

79. *Journal of Pediatrics* 45: 179 (1954).

80. *Indian Journal of Medical Research* 56 (1): 12 (1968).

81. S. Cleary, ed., *Biological Effects and Health Implications of Microwave Radiations*, U.S. Public Health Service Publication PB 193: 898 (1969).

82. *Thrombosis et Diathesis Haemorphagica* 26 (3): 417 (1971).

83. *Archives of Internal Medicine* 131: 688 (1973).

84. G. Hanneman and J. Sershon, "Tolerance Endpoint for Evaluating the Effects of Heat Stress in Dogs," FAA Report #FAA-AM-84–5, June, 1984.

85. *Journal of Applied Physiology* 53: 1171–1174 (1982).

86. *Aviation, Space and Environmental Medicine* 57: 659–663 (1986).

87. B. Zweifach, "Aspects of Comparative Physiology of Laboratory Animals Relative to Problems of Experimental Shock," Federal Procedings 20, Suppl. 9: 18–29 (1961); cited in *Aviation, Space and Environmental Medicine* 50 (8):8–19 (1979).

88. *Annual Review of Physiology* 8: 335 (1946).

89. *Pharmacological Review* 6 (4): 489 (1954).

90. K. Hobler and R. Napodano, *Journal of Trauma* 14 (8): 716 (1974).

91. Martin Stephens, *A Critique of Animal Experiments on Cocaine Abuse*, a report prepared for the American, National and New England Anti-Vivisection Societies (Boston, 1985).

92. *Health Care* 2 (26), August 28–September 10, 1980.

93. *Journal of Pharmacology and Experimental Therapeutics* 226 (3): 783–789 (1983).

94. *Psychopharmacology* 88: 500–504 (1986).

95. *Bulletin of the Psychonomic Society* 22 (1): 53–56 (1984).

96. *European Journal of Pharmacology* 40: 114–115 (1976).

97. *Newsweek*, December 26, 1988, p. 50; "TSU Shuts Down Cornell Cat Lab," *The Animals' Agenda* 9 (3): 22–25 (March 1989).

98. S. Milgram, *Obedience to Authority* (New York: Harper and Row, 1974). Incidentally, these experiments were widely criticized on ethical grounds because they involved human beings without their consent. It is indeed questionable whether Milgram should have deceived participants in his experiment as he did; but when we compare what was done to them with what is commonly done to nonhuman animals, we can appreciate how far more sensitive most people are when assessing the ethics of experimentation with humans.

99. *Monitor*, a publication of the American Psychological Association, March 1978.

100. Donald J. Barnes, "A Matter of Change," in Peter Singer, ed., *In Defense of Animals*, pp. 160, 166.

101. *The Death Sciences in Veterinary Research and Education*, (New York: United Action for Animals), p. iii.

102. *Journal of the American Veterinary Medical Association* 163 (9): 1 (November 1, 1973).

103. See Appendix 3 for the address.

104. *Journal of Comparative and Physiological Psychology* 55: 896 (1962).

105. *Scope* (Durban, South Africa), March 30, 1973.

106. Robert J. White, "Antivivisection: The Reluctant Hydra," *The American Scholar* 40 (1971); reprinted in T. Regan and P. Singer, eds., *Animal Rights and Human Obligations*, first edition, (Englewood Cliffs, N.J.: Prentice-Hall, 1976), p. 169.

107. *The Plain Dealer*, July 3, 1988.

108. *Birmingham News*, Birmingham, Alabama, February 12, 1988.

109. "The Price of Knowledge," broadcast in New York, December 12, 1974, WNET/13, transcript supplied courtesy of WNET/13 and Henry Spira.

110. Quoted in the OTA report, *Alternatives to Animal Use in Research, Testing and Education*, p. 277.

111. National Health and Medical Research Council, *Code of Practice for the Care and Use of Animals for Experimental Purposes* (Australian Government Publishing Service, Canberra, 1985). A revised code has recently been accepted; see "Australian Code of Practice," *Nature* 339: 412 (June 8, 1989).

112. OTA, *Alternatives to Animal Use in Research, Testing and Education*, p. 377.

113. Pat Monaghan, "The Use of Animals in Medical Research," *New Scientist*, November 19, 1988, p. 54.

114. For a summary of the 1985 amendments, and of the law and regulations at this time, see OTA, *Alternatives to Animal Use in Research, Testing and Education*, pp. 280–286.

115. OTA, *Alternatives to Animal Use in Research, Testing, and Education*, pp. 286–287.

116. OTA, *Alternatives to Animal Use in Research, Testing, and Education*, pp.287, 298.

117. National Research Council, *Use of Laboratory Animals in Biomedical and Behavioral Research* (Washington, D.C.: National Academy Press, 1988). See especially the "Individual Statement" by C. Stevens.

118. *The Washington Post*, July 19, 1985, p. A10. For a more detailed account of the Gennarelli case, see Lori Gruen and Peter Singer, *Animal Liberation: A Graphic Guide* (London: Camden Press, 1987), pp. 10–23.

119. "Group Charges Gillette Abuses Lab Animals," *Chemical and Engineering News*, October 6, 1986, p. 5.

120. H. Beecher, "Ethics and Clinical Research," *New England Journal of Medicine* 274: 1354–1360 (1966); D. Rothman, "Ethics and Human Experimentation: Henry Beecher Revisited," *New England Journal of Medicine* 317: 1195–1199 (1987).

121. From the transcript of the "Doctors Trial," Case I, United States vs. Brandt et al. Quoted by W. L. Shirer, *The Rise and Fall of the Third Reich* (New York: Simon and Schuster, 1960), p. 985. For additional descriptions of these experiments, see R. J. Lifton, *The Nazi Doctors* (New York: Basic Books, 1986).

122. *British Journal of Experimental Pathology* 61: 39 (1980); cited by R. Ryder, "Speciesism in the Laboratory," in Peter Singer, ed., *In Defense of Animals*, p. 85.

123. I. B. Singer, *Enemies: A Love Story* (New York: Farrar, Straus and Giroux, 1972).

124. See James Jones, *Bad Blood: The Tuskegee Syphilis Experiment* (New York: Free Press, 1981).

125. Sandra Coney, *The Unfortunate Experiment* (Auckland: Penguin Books, 1988).

126. E. Wynder and D. Hoffman, in *Advances in Cancer Research* 8 (1964); see also the Royal College of Physicians report, *Smoking and Health* (London, 1962) and studies by the U.S. Health Depart-

ment. I owe these references to Richard Ryder, "Experiments on Animals," in Stanley and Roslind Godlovitch and John Harris, eds., *Animals, Men and Morals*, p. 78.

127. "U.S. Lung Cancer Epidemic Abating, Death Rates Show," *The Washington Post*, October 18, 1989, p. 1.

128. "The Cancer Watch," *U.S. News & World Report*, February 15, 1988.

129. *Science* 241: 79 (1988).

130. "Colombians Develop Effective Malaria Vaccine," *The Washington Post*, March 10, 1988.

131. "Vaccine Produces AIDS Antibodies," *Washington Times*, April 19, 1988.

132. "AIDS Policy in the Making," *Science* 239: 1087 (1988).

133. T. McKeown, *The Role of Medicine: Dream, Mirage or Nemesis?* (Oxford: Blackwell, 1979).

134. D. St. George, "Life Expectancy, Truth, and the ABPI," *The Lancet*, August 9, 1986, p. 346.

135. J. B. McKinlay, S. M. McKinlay and R. Beaglehole, "Trends in Death and Disease and the Contribution of Medical Measures" in H. E. Freeman and S. Levine, eds., *Handbook of Medical Sociology* (Englewood Cliffs, N.J.: Prentice Hall, 1988), p. 16.

136. See William Paton, *Man and Mouse* (Oxford: Oxford University Press, 1984); Andrew Rowan, *Of Mice, Models and Men: A Critical Evaluation of Animal Research* (Albany: State University of New York Press, 1984), chapter 12; Michael DeBakey, "Medical Advances Resulting From Animal Research," in J. Archibald, J. Ditchfield, and H. Rowsell, eds., *The Contribution of Laboratory Animal Science to the Welfare of Man and Animals: Past, Present and Future* (New York: Gustav Fischer Verlag, 1985); OTA, *Alternatives to Animal Use in Research, Testing and Education*, chapter 5;

and National Research Council, *Use of Animals in Biomedical and Behavioral Research* (National Academy Press, Washington, D.C., 1988), Chapter 3.

137. Probably the best of those works arguing against the claims made for animal experimentation is Robert Sharpe, *The Cruel Deception*.

138. "The Costs of AIDS," *New Scientist*, March 17, 1988, p. 22.

Chapter 3

1. *The Washington Post*, October 3, 1971; see also the testimony during September and October 1971, before the Subcommittee on Monopoly of the Select Committee on Small Business of the U.S. Senate, in the Hearings on the Role of Giant Corporations, especially the testimony of Jim Hightower of the Agribusiness Accountability Project. For the size of egg producers, see *Poultry Tribune*, June 1987, p. 27.

2. Ruth Harrison, *Animal Machines* (London: Vincent Stuart, 1964), p. 3.

3. *Broiler Industry*, December 1987, p. 22.

4. Konrad Lorenz, *King Solomon's Ring* (London: Methuen and Company, 1964), p. 147.

5. *Farming Express*, February 1, 1962; quoted by Ruth Harrison, *Animal Machines*, p. 18.

6. F. D. Thornberry, W. O. Crawley, and W. F. Krueger, "Debeaking: Laying Stock to Control Cannibalism," *Poultry Digest*, May 1975, p. 205.

7. As reported in *The Animal Welfare Institute Quarterly*, Fall 1987, p. 18.

8. *Report of the Technical Committee to Enquire into the Welfare of Animals Kept Under Intensive Livestock Husbandry Systems*, Command Paper 2836 (London: Her Majesty's Stationery Office, 1965), paragraph 97.

9. A. Andrade and J. Carson, "The Effect of Age and Methods of Debeaking on Future Performance of White Leghorn Pullets," *Poultry Science* 54: 666–674 (1975); M. Gentle, B. Huges, and R. Hubrecht, "The Effect of Beak Trimming on Food Intake, Feeding Behavior and Body Weight in Adult Hens," *Applied Animal Ethology* 8: 147–159 (1982); M. Gentle, "Beak Trimming in Poultry," *World's Poultry Science Journal* 42: 268–275 (1986).

10. J. Breward and M. Gentle, "Neuroma Formation and Abnormal Afferent Nerve Discharges After Partial Beak Amputation (Beak Trimming) in Poultry," *Experienta* 41: 1132–1134 (1985).

11. Gentle, "Beak Trimming in Poultry," *World's Poultry Science Journal* 42: 268–275 (1986).

12. U.S. Department of Agriculture Yearbook for 1970, p. xxxiii.

13. *Poultry World*, December 5, 1985.

14. *American Agriculturist*, March 1967.

15. C. Riddell and R. Springer, "An Epizootiological Study of Acute Death Syndrome and Leg Weakness in Broiler Chickens in Western Canada," *Avian Diseases* 29: 90–102 (1986); P. Steele and J. Edgar, "Importance of Acute Death Syndrome in Mortalities in Broiler Chicken Flocks," *Poultry Science* 61: 63–66 (1982).

16. R. Newberry, J. Hunt, and E. Gardiner, "Light Intensity Effects on Performance, Activity, Leg Disorders, and Sudden Death Syndrome of Roaster Chickens, *Poultry Science* 66: 1446–1450 (1987).

17. Trevor Bray, as reported in *Poultry World*, June 14, 1984.

18. See the studies by Riddell and Springer, and by Steele and Edgar, cited in note 15 above.

19. D. Wise and A. Jennings, "Dyschondroplasia in Domestic Poultry," *Veterinary Record* 91: 285–286 (1972).

20. G. Carpenter et al., "Effect of Internal Air Filtration on the Performance of Broilers and the Aerial Concentrations of Dust and Bacteria," *British Poultry Journal* 27: 471–480 (1986).

21. "Air in Your Shed a Risk to Your Health," *Poultry Digest*, December/January 1988.

22. *The Washington Times*, October 22, 1987.

23. *Broiler Industry*, December 1987, and *Hippocrates*, September/October 1988. Perdue has confirmed that his chickens are debeaked in a letter to me. See also the Animal Rights International advertisement, "Frank, are you telling the truth about your chickens?" *The New York Times*, October 20, 1989, p. A17.

24. F. Proudfoot, H. Hulan, and D. Ramey, "The Effect of Four Stocking Densities on Broiler Carcass Grade, the Incidence of Breast Blisters, and Other Performance Traits," *Poultry Science* 58: 791–793 (1979).

25. *Turkey World*, November/December 1986.

26. *Poultry Tribune*, January 1974.

27. *Farmer and Stockbreeder*, January 30, 1982; quoted by Ruth Harrison, *Animal Machines*, p. 50.

28. *Feedstuffs*, July 25, 1983.

29. *American Agriculturist*, July 1966.

30. USDA statistics indicate that in 1986 the population of commercial layers was 246 million. Assuming that the male/female hatching ratio is roughly 50 percent, and that each bird is re-

placed approximately every eighteen months, the above estimate is a minimum one.

31. *American Agriculturist*, March 1967.

32. *Upstate*, August 5, 1973, report by Mary Rita Kiereck.

33. *National Geographic Magazine*, February 1970.

34. *Poultry Tribune*, February 1974.

35. *Federal Register*, December 24, 1971, p. 24926.

36. *Poultry Tribune*, November 1986.

37. First report from the Agriculture Committee, House of Commons, 1980–1981, *Animal Welfare in Poultry, Pig and Veal Production* (London: Her Majesty's Stationery Office, 1981), paragraph 150.

38. B. M. Freeman, "Floor Space Allowance for the Caged Domestic Fowl," *The Veterinary Record*, June 11, 1983, pp. 562–563.

39. *Poultry Tribune*, March 1987, p. 30; "Swiss Federal Regulations on Animal Protection," May 29, 1981.

40. Information on the Netherlands provided by Compassion in World Farming, and the Netherlands Embassy, London. (See also *Farmer's Guardian.*, September 29, 1989.) On Sweden, see Steve Lohr, "Swedish Farm Animals Get a New Bill of Rights," *The New York Times*, October 25, 1988.

41. *Poultry Tribune*, March 1987.

42. European Parliament, Session 1986/7, Minutes of Proceedings of the Sitting of 20 February 1987, Document A2-211/86.

43. *Poultry Tribune*, November 1986.

44. *Upstate*, August 5, 1973.

45. *Animal Liberation (Victoria) Newsletter*, May 1988 and February 1989.

46. Roy Bedichek, *Adventures with a Naturalist*, quoted by Ruth Harrison, *Animal Machines*, p. 154.

47. *Upstate*, August 5, 1973.

48. *Der Spiegel*, 1980, no. 47, p. 264; quoted in *Intensive Egg and Chicken Production*, Chickens' Lib (Huddersfield, U.K.)

49. I. Duncan and V. Kite, "Some Investigations into Motivation in the Domestic Fowl," *Applied Animal Behaviour Science* 18: 387–388 (1987).

50. *New Scientist*, January 30, 1986, p. 33, reporting a study by H. Huber, D. Fölsch, and U. Stahli, published in *British Poultry Science* 26: 367 (1985).

51. A. Black and B. Hughes, "Patterns of Comfort Behaviour and Activity in Domestic Fowls: A Comparison Between Cages and Pens," *British Veterinary Journal* 130: 23–33 (1974).

52. D. van Liere and S. Bokma, "Short-term Feather Maintenance as a Function of Dust-bathing in Laying Hens," *Applied Animal Behaviour Science* 18: 197–204 (1987).

53. H. Simonsen, K. Vestergaard, and P. Willeberg, "Effect of Floor Type and Density on the Integument of Egg Layers," *Poultry Science* 59: 2202–2206 (1980).

54. K. Vestergaard, "Dustbathing in the Domestic Fowl—Diurnal Rhythm and Dust Deprivation," *Applied Animal Behaviour Science* 17: 380 (1987).

55. H. Simonsen, K. Vestergaard, and P. Willeberg, "Effect of Floor Type and Density on the Integument of Egg Layers."

56. J. Bareham, "A Comparison of the Behaviour and Production of Laying Hens in Experimental and Conventional Battery

Cages," *Applied Animal Ethology* 2: 291–303 (1976).

57. J. Craig, T. Craig, and A. Dayton, "Fearful Behavior by Caged Hens of Two Genetic Stocks," *Applied Animal Ethology* 10: 263–273 (1983).

58. M. Dawkins, "Do Hens Suffer in Battery Cages? Environmental Preferences and Welfare," *Applied Animal Behaviour* 25: 1034–1046 (1977). See also M. Dawkins, *Animal Suffering: The Science of Animal Welfare* (London: Chapman and Hall, 1980), chapter 7.

59. *Plain Truth* (Pasadena, California), March 1973.

60. C. E. Ostrander and R. J. Young, "Effects of Density on Caged Layers," *New York Food and Life Sciences* 3 (3) (1970).

61. U.K. Ministry of Agriculture, Fisheries and Food, Poultry Technical Information Booklet No. 13; quoted from *Intensive Egg and Chicken Production*, Chickens' Lib (Huddersfield, U.K.).

62. *Poultry Tribune*, March 1974.

63. Ian Duncan, "Can the Psychologist Measure Stress?" *New Scientist*, October 18, 1973.

64. R. Dunbar, "Farming Fit for Animals," *New Scientist*, March 29, 1984, pp. 12–15; D. Wood-Gush, "The Attainment of Humane Housing for Farm Livestock," in M. Fox and L. Mickley, eds., *Advances in Animal Welfare Science* (Washington, D.C.: Humane Society of the United States, 1985).

65. *Farmer's Weekly*, November 7, 1961, quoted by Ruth Harrison, *Animal Machines*, p. 97.

66. R. Dantzer and P. Mormede, "Stress in Farm Animals: A Need for Reevaluation," *Journal of Animal Science* 57: 6–18 (1983).

67. D. Wood-Gush and R. Beilharz, "The Enrichment of a Bare Environment for Animals in Confined Conditions," *Applied Ani-*

mal Ethology 20: 209–217 (1983).

68. U.S. Department of Agriculture, Fact Sheet: Swine Management, AFS-3-8-12, Department of Agriculture, Office of Governmental and Public Affairs, Washington, D.C.

69. F. Butler, quoted in John Robbins, *Diet for a New America* (Walpole, N.H.: Stillpoint, 1987). p. 90.

70. D. Fraser, "The Role of Behaviour in Swine Production: a Review of Research," *Applied Animal Ethology* 11: 332 (1984).

71. D. Fraser, "Attraction to Blood as a Factor in Tail Biting by Pigs," *Applied Animal Behaviour Science* 17: 61–68 (1987).

72. *Farm Journal*, May 1974.

73. The relevant studies are summarized by Michael W. Fox, *Farm Animals: Husbandry, Behavior, Veterinary Practice* (University Park Press, 1984), p. 126.

74. *Farmer and Stockbreeder* January 22, 1963; quoted by Ruth Harrison, *Animal Machines*, p. 95.

75. "Swine Production Management," Hubbard Milling Company, Mankato, Minnesota, 1984.

76. William Robbins, "Down on the Superfarm: Bigger Share of Profits," *The New York Times*, August 4, 1987.

77. *Feedstuffs*, January 6, 1986, p. 6.

78. *Hog Farm Management*, December 1975, p. 16.

79. Bob Frase, quoted in Orville Schell, *Modern Meat* (New York: Random House, 1984), p. 62.

80. *Farmer and Stockbreeder*, July 11, 1961; quoted in Ruth Harrison, *Animal Machines*, p. 148.

81. J. Messersmith, quoted in J. Robbins, *Diet for a New America*, p. 84.

82. *Agscene* (Petersfield, Hampshire, England), June 1987, p. 9.

83. *Farm Journal*, March 1973.

84. "Mechanical Sow Keeps Hungry Piglets Happy," *The Western Producer*, April 11, 1985.

85. *National Hog Farmer*, March 1978, p. 27.

86. U.S. Department of Agriculture, Fact Sheet: Swine Management, AFS-3-8-12, Department of Agriculture, Office of Governmental and Public Affairs, Washington, D.C.

87. U.S. Department of Agriculture, Fact Sheet: Swine Housing, AFS-3-8-9, Department of Agriculture, Office of Governmental and Public Affairs, Washington, D.C.

88. G. Cronin, "The Development and Significance of Abnormal Stereotyped Behaviour in Tethered Sows," Ph.D. thesis, University of Wageningen, Netherlands, p. 25.

89. Roger Ewbank, "The Trouble with Being a Farm Animal," *New Scientist*, October 18, 1973.

90. "Does Close Confinement Cause Distress in Sows?" Scottish Farm Buildings Investigation Unit, Aberdeen, July 1986, p. 6.

91. Farm Animal Welfare Council, *Assessment of Pig Production Systems* (Farm Animal Welfare Council, Surbiton, Surrey, England, 1988), p. 6.

92. A. Lawrence, M. Appleby, and H. MacLeod, "Measuring Hunger in the Pig Using Operant Conditioning: The Effect of Food Restriction," *Animal Production* 47 (1988).

93. *The Stall Street Journal*, July 1972.

94. J. Webster, C. Saville, and D. Welchman, "Improved Husbandry Systems for Veal Calves," Animal Health Trust and Farm Animal Care Trust, no date, p. 5; see also Webster et al., "The Effect of Different Rearing Systems on the Development of Calf Behavior," and "Some Effects of Different Rearing Systems on Health, Cleanliness and Injury in Calves," *British Veterinary Journal* 1141: 249 and 472 (1985).

95. J. Webster, C. Saville, and D. Welchman, "Improved Husbandry Systems for Veal Calves," p. 6.

96. J. Webster, C. Saville, and D. Welchman, "Improved Husbandry Systems for Veal Calves," p. 2.

97. *The Stall Street Journal*, November 1973.

98. *The Stall Street Journal*, April 1973.

99. *The Stall Street Journal*, November 1973.

100. *Farmer and Stockbreeder*, September 13, 1960, quoted by Ruth Harrison, *Animal Machines*, p. 70.

101. *The Stall Street Journal*, April 1973.

102. G. van Putten, "Some General Remarks Concerning Farm Animal Welfare in Intensive Farming Systems," unpublished paper from the Research Institute for Animal Husbandry, "Schoonoord," Driebergseweg, Zeist, The Netherlands, p. 2.

103. G. van Putten, "Some General Remarks Concerning Farm Animal Welfare in Intensive Farming Systems," p. 3.

104. *The Vealer*, March/April 1982.

105. U.K. Ministry of Agriculture, Fisheries and Food, Welfare of Calves Regulations, 1987 (London: Her Majesty's Stationery Office, 1987).

106. J. Webster, "Health and Welfare of Animals in Modern Hus-

bandry Systems—Dairy Cattle," *In Practice*, May 1986, p. 85.

107. Gordon Harvey, "Poor Cow," *New Scientist*, September 29, 1983, pp. 940–943.

108. *The Washington Post*, March 28, 1988.

109. D. S. Kronfeld, "Biologic and Economic Risks Associated with Bovine Growth Hormone," Conference on Growth Hormones, European Parliament, December 9, 1987, unpublished paper, p. 4.

110. D. S. Kronfeld, "Biologic and Economic Risks Associated with Bovine Growth Hormone," p. 5.

111. Bob Holmes, "Secrecy Over Cow Hormone Experiments," *Western Morning News*, January 14, 1988.

112. Keith Schneider, "Better Farm Animals Duplicated by Cloning," *The New York Times*, February 17, 1988; see also Ian Wilmut, John Clark, and Paul Simons, "A Revolution in Animal Breeding," *New Scientist*, July 7, 1988.

113. *The Peoria Journal Star*, June 5, 1988.

114. "Is Pain the Price of Farm Efficiency?" *New Scientist*, October 13, 1973, p.171.

115. *Feedstuffs*, April 6, 1987.

116. *Farm Journal*, August 1967, March 1968.

117. S. Lukefahr, D. Caveny, P. R. Cheeke, and N. M. Patton, "Rearing Weanling Rabbits in Large Cages," *The Rabbit Rancher*, cited in Australian Federation of Animal Societies, *Submission to the Senate Select Committee of Inquiry into Animal Welfare in Australia*, vol. 2, Melbourne, 1984.

118. *The Age* (Melbourne), May 25, 1985.

119. This cage size is recommended by the Finnish Fur Breeders Association. For mink, the U.K. Fur Breeders Association recommends cages thirty inches by nine inches. See Fur Trade Fact Sheet, Lynx (1986), Great Dunmow, Essex.

120. *Report of the Technical Committee to Enquire into the Welfare of Animals Kept Under Intensive Livestock Husbandry Systems*, appendix.

121. *Report of the Technical Committee to Enquire into the Welfare of Animals Kept Under Intensive Livestock Husbandry Systems*, paragraph 37.

122. See page 120 above.

123. Joy Mensch and Ari van Tienhove, "Farm Animal Welfare," *American Scientist*, November/December 1986, p. 599, citing a paper by D. W. Fölsch, "Egg Production—Not Necessarily a Reliable Indicator for the State of Health of Injured Hens," in *Fifth European Poultry Conference*, Malta, 1976.

124. B. Gee, *The 1985 Muresk Lecture*, Muresk Agricultural College, Western Australian Institute of Technology, p. 8.

125. European Parliament, Session 1986/7, Minutes of proceedings of the sitting of February 20, 1987, Document A2-211/86.

126. D. W. Fölsch, et al, "Research on Alternatives to the Battery System for Laying Eggs," *Applied Animal Behaviour Science* 20: 29–45 (1988).

127. *Dehorning, Castrating, Branding, Vaccinating Cattle*, Publication No. 384 of the Mississippi State University Extension Service, cooperating with the USDA; see also *Beef Cattle: Dehorning, Castrating, Branding and Marking*, USDA, Farmers' Bulletin No. 2141, September 1972.

128. *Progressive Farmer*, February 1969.

129. *Pig Farming*, September 1973.

130. *Hot-iron Branding*, University of Georgia College of Agriculture, Circular 551.

131. *Beef Cattle: Dehorning, Castrating, Branding and Marking*.

132. R. F. Bristol, "Preconditioning of Feeder Cattle Prior to Interstate Shipment," Report of a Preconditioning Seminar held at Oklahoma State University, September 1967, p. 65.

133. U.S. Department of Agriculture Statistical Summary, Federal Meat and Poultry Inspection for Fiscal Year 1986.

134. *The Washington Post*, September 30, 1987.

135. Colman McCarthy, "Those Who Eat Meat Share in the Guilt," *The Washington Post*, April 16, 1988.

136. Farm Animal Welfare Council, *Report on the Welfare of Livestock (Red Meat Animals) at the Time of Slaughter* (London: Her Majesty's Stationery Office, 1984) paragraphs 88, 124.

137. Harold Hillman, "Death by Electricity," *The Observer* (London) July 9, 1989.

138. "Animals into Meat: A Report on the Pre-Slaughter Handling of Livestock," Argus Archives (New York) 2: 16–17 (March 1970); the description is by John MacFarlane, a vice president of Livestock Conservation, Inc.

139. Farm Animal Welfare Council, *Report on the Welfare of Livestock When Slaughtered by Religious Methods* (London: Her Majesty's Stationery Office, 1985), paragraph 50.

140. Temple Grandin, letter dated November 7, 1988.

141. Farm Animal Welfare Council, *Report on the Welfare of Livestock When Slaughtered by Religious Methods*, paragraph 27.

142. *Science* 240: 718 (May 6, 1988).

143. Caroline Murphy, "The 'New Genetics' and the Welfare of Animals," *New Scientist*, December 10, 1988, p. 20.

144. "Genetic Juggling Raises Concerns," *The Washington Times*, March 30, 1988.

Chapter 4

1. Oliver Goldsmith, *The Citizen of the World*, in *Collected Works*, ed. A. Friedman (Oxford: Clarendon Press, 1966), vol. 2, p. 60. Apparently Goldsmith himself fell into this category, however, since according to Howard Williams in *The Ethics of Diet* (abridged edition, Manchester and London, 1907, p. 149), Goldsmith's sensibility was stronger than his self-control.

2. In attempting to rebut the argument for vegetarianism presented in this chapter of the first edition, R. G. Frey described the reforms proposed by the House of Commons Agriculture Committee in 1981, and wrote: "The House of Commons as a whole has yet to decide on this report, and it may well be in the end that it is diluted; but even so, there is no doubt that it represents a significant advance in combating the abuses of factory farming." Frey then argued that the report showed that these abuses could be overcome by tactics that stopped short of advocating a boycott of animal products. (R. G. Frey, *Rights, Killing and Suffering*, Oxford: Blackwell, 1983, p. 207.) This is one instance in which I sincerely wish my critic had been right; but the House of Commons did not bother to "dilute" the report of its Agriculture Committee—it simply ignored it. Eight years later nothing has changed for the vast majority of Britain's intensively produced animals. The exception is for veal calves, where a consumer boycott *did* play a significant role.

3. Frances Moore Lappé, *Diet for a Small Planet* (New York: Friends of the Earth/Ballantine, 1971), pp. 4–11. This book is the best brief introduction to the topic, and figures not otherwise attributed in this section have been taken from this book. (A re-

vised edition was published in 1982.) The main original sources are *The World Food Problem,* a Report of the President's Science Advisory Committee (1967); *Feed Situation,* February 1970, U.S. Department of Agriculture; and *National and State Livestock-Feed Relationships,* U.S. Department of Agriculture, Economic Research Service, Statistical Bulletin No. 446, February 1970.

4. The higher ratio is from Folke Dovring, "Soybeans," *Scientific American,* Feburary 1974. Keith Akers presents a different set of figures in *A Vegetarian Sourcebook* (New York: Putnam, 1983), chapter 10. His tables compare per-acre nutritional returns from oats, broccoli, pork, milk, poultry, and beef. Even though oats and broccoli are not high-protein foods, none of the animal foods produced even half as much protein as the plant foods. Akers's original sources are: United States Department of Agriculture, *Agricultural Statistics,* 1979; United States Department of Agriculture, *Nutritive Value of American Foods* (Washington, D.C., U.S. Government Printing Office, 1975); and C. W. Cook, "Use of Rangelands for Future Meat Production," *Journal of Animal Science* 45: 1476 (1977).

5. Keith Akers, *A Vegetarian Sourcebook,* pp. 90–91, using the sources cited above.

6. Boyce Rensberger, "Curb on U.S. Waste Urged to Help World's Hungry," *The New York Times,* October 25, 1974.

7. *Science News,* March 5, 1988, p. 153, citing *Worldwatch,* January/February 1988.

8. Keith Akers, *A Vegetarian Sourcebook,* p. 100, based on D. Pimental and M. Pimental, *Food, Energy and Society* (New York: Wiley, 1979), pp. 56, 59, and U.S. Department of Agriculture, *Nutritive Value of American Foods* (Washington, D.C.: U.S. Government Printing Office, 1975).

9. G. Borgstrom, *Harvesting the Earth* (New York: Abelard-Schuman, 1973) pp. 64–65; cited by Keith Akers, *A Vegetarian Sourcebook.*

10. "The Browning of America," *Newsweek*, February 22, 1981, p. 26; quoted by John Robbins, *Diet for a New America* (Walpole, N.H.: Stillpoint, 1987), p. 367.

11. "The Browning of America," p. 26.

12. Fred Pearce, "A Green Unpleasant Land," *New Scientist*, July 24, 1986, p. 26.

13. Sue Armstrong, "Marooned in a Mountain of Manure," *New Scientist*, November 26, 1988.

14. J. Mason and P. Singer, *Animal Factories* (New York: Crown, 1980), p. 84, citing R. C. Loehr, *Pollution Implications of Animal Wastes—A Forward Oriented Review*, Water Pollution Control Research Series (U.S. Environmental Protection Agency, Washington, D.C., 1968), pp. 26–27; H. A. Jasiorowski, "Intensive Systems of Animal Production," in R. L. Reid, ed., *Proceedings of the II World Conference on Animal Production* (Sydney: Sydney University Press, 1975), p. 384; and J. W. Robbins, *Environmental Impact Resulting from Unconfined Animal Production* (Cincinnati: Environmental Research Information Center, U.S. Environmental Protection Agency, 1978) p. 9.

15. "Handling Waste Disposal Problems," *Hog Farm Management*, April 1978, p. 17, quoted in J. Mason and P. Singer, *Animal Factories*, p. 88.

16. Information from the Rainforest Action Network, *The New York Times*, January 22, 1986, p. 7.

17. E. O. Williams, *Biophilia* (Cambridge: Harvard University Press, 1984), p. 137.

18. Keith Akers, *A Vegetarian Sourcebook*, pp. 99–100; based on H. W. Anderson, et al., *Forests and Water: Effects of Forest Management on Floods, Sedimentation and Water Supply*, U.S. Department of Agriculture Forest Service General Technical Report PSW–18/1976; and J. Kittridge, "The Influence of the Forest on the Weather and other Environmental Factors," in United Nations

Food and Agriculture Organization, *Forest Influences* (Rome, 1962).

19. Fred Pearce, "Planting Trees for a Cooler World," *New Scientist*, October 15, 1988, p. 21.

20. David Dickson, "Poor Countries Need Help to Adapt to Rising Sea Level," *New Scientist*, October 7, 1989, p. 4; Sue Wells and Alasdair Edwards, Gone with the Waves," *New Scientist*, November 11, 1989, pp. 29–32.

21. L. and M. Milne, *The Senses of Men and Animals* (Middlesex and Baltimore: Penguin Books, 1965), chapter 5.

22. *Report of the Panel of Enquiry into Shooting and Angling*, published by the panel in 1980 and available through the Royal Society for the Prevention of Cruelty to Animals (U.K.), paragraphs 15–57.

23. Geoff Maslen, "Bluefin, the Making of the Mariners," *The Age* (Melbourne), January 26, 1985.

24. D. Pimental and M. Pimental, *Food, Energy and Society* (New York: Wiley, 1979), chapter 9; I owe this reference to Keith Akers, *A Vegetarian Sourcebook*, p. 117.

25. See J. R. Baker: *The Humane Killing of Lobsters and Crabs*, The Humane Education Centre, London, no date; J. R. Baker and M. B. Dolan, "Experiments on the Humane Killing of Lobsters and Crabs," *Scientific Papers of the Humane Education Centre* 2: 1–24 (1977).

26. My change of mind about mollusks stems from conversations with R. I. Sikora.

27. See pages 230–231 below.

28. "Struggle" is not altogether a joke. According to a comparative study published in *The Lancet* (December 30, 1972), the "mean transit time" of food through the digestive system of a

sample group of nonvegetarians on a Western type of diet was between seventy-six and eighty-three hours; for vegetarians, forty-two hours. The authors suggest a connection between the length of time the stool remains in the colon and the incidence of cancer of the colon and related diseases which have increased rapidly in nations whose consumption of meat has increased but are almost unknown among rural Africans who, like vegetarians, have a diet low in meat and high in roughage.

29. David Davies, "A Shangri-La in Ecuador," *New Scientist*, February 1, 1973. On the basis of other studies, Ralph Nelson of the Mayo Medical School has suggested that a high protein intake causes us to "idle our metabolic engine at a faster rate" (*Medical World News*, November 8, 1974, p. 106). This could explain the correlation between longevity and little or no meat consumption.

30. *The Surgeon General's Report on Nutrition and Health* (Washington, D.C.: U.S. Government Printing Office, 1988).

31. According to a wire-service report cited in *Vegetarian Times*, November 1988.

32. *The New York Times*, October 25, 1974.

33. N. Pritikin and P. McGrady, *The Pritikin Program for Diet and Exercise* (New York: Bantam, 1980); J. J. McDougall, *The McDougall Plan* (Piscataway, N.J.: New Century, 1983).

34. Francis Moore Lappé, *Diet for a Small Planet*, pp. 28–29; see also *The New York Times*, October 25, 1974; *Medical World News*, November 8, 1974, p. 106.

35. Quoted in F. Wokes, "Proteins," *Plant Foods for Human Nutrition*, 1: 38 (1968).

36. In the first edition of *Diet for a Small Planet* (1971), Frances Moore Lappé emphasized protein complementarity to show that a vegetarian diet can provide enough protein. In the revised edition (New York: Ballantine, 1982) this emphasis has disappeared, replaced by a demonstration that a healthy vegetarian

diet is bound to contain enough protein even without complementarity. For another account of the adequacy of plant foods as far as protein is concerned, see Keith Akers, *A Vegetarian Sourcebook,* chapter 2.

37. F. R. Ellis and W. M. E. Montegriffo, "The Health of Vegans," *Plant Foods for Human Nutrition,* vol. 2, pp. 93-101 (1971). Some vegans claim that B12 supplements are not necessary, on the grounds that the human intestine can synthesize this vitamin from other B-group vitamins. The question is, however, whether this synthesis takes place sufficiently early in the digestive process for the B12 to be absorbed rather than excreted. At present the nutritional adequacy of an all-plant diet without supplementation is an open scientific question; accordingly it seems safer to take supplementary B12. See also F. Wokes, "Proteins," *Plant Foods for Human Nutrition,* p. 37.

Chapter 5

1. Genesis 1:24–28.

2. Genesis 9:1–3.

3. *Politics,* Everyman's Library, (London: J. M. Dent & Sons, 1959), p. 10.

4. *Politics,* p. 16.

5. W. E. H. Lecky, *History of European Morals from Augustus to Charlemagne* (London: Longmans, 1869), 1: 280–282.

6. Mark 5:1–13.

7. Corinthians 9:9–10.

8. Saint Augustine, *The Catholic and Manichaean Ways of Life,* trans. D. A. Gallagher and I. J. Gallagher (Boston: The Catholic

University Press, 1966), p. 102. I owe this reference to John Passmore, *Man's Responsibility for Nature* (New York: Scribner's, 1974), p. 11.

9. *History of European Morals*, vol. 1, p. 244; for Plutarch see especially the essay "On the Eating of Flesh" in his *Moral Essays*.

10. For Basil, see John Passmore, "The Treatment of Animals," *The Journal of the History of Ideas* 36: 198 (1975); for Chrysostom, Andrew Linzey, *Animal Rights: A Christian Assessment of Man's Treatment of Animals* (London: SCM Press, 1976), p. 103; and for Saint Isaac the Syrian, A. M. Allchin, *The World is a Wedding: Explorations in Christian Spirituality* (London: Darton, Longman and Todd, 1978), p. 85. I owe these references to R. Attfield, "Western Traditions and Environmental Ethics," in R. Elliot and A. Gare, eds., *Environmental Philosophy* (St. Lucia: University of Queensland Press, 1983), pp. 201–230. For further discussion see Attfield's book, *The Ethics of Environmental Concern* (Oxford: Blackwell, 1982); K. Thomas, *Man and the Natural World: Changing Attitudes in England 1500–1800* (London: Allen Lane, 1983), pp. 152–153; and R. Ryder, *Animal Revolution: Changing Attitudes Towards Speciecism* (Oxford: Blackwell, 1989), pp 34–35.

11. *Summa Theologica* II, II, Q64, art. 1.

12. *Summa Theologica* II, II, Q159, art. 2.

13. *Summa Theologica* I, II, Q72, art. 4.

14. *Summa Theologica* II, II, Q25, art. 3.

15. *Summa Theologica* II, I, Q102, art. 6; see also *Summa contra Gentiles* III, II, 112 for a similar view.

16. E. S. Turner, *All Heaven in Rage* (London: Michael Joseph, 1964), p. 163.

17. V. J. Bourke, *Ethics* (New York: Macmillan, 1951), p. 352.

18. John Paul II, *Solicitudo Rei Socialis* (Homebush, NSW: St. Paul Publications, 1988), sec. 34, pp. 73–74.

19. *St. Francis of Assisi, His Life and Writings as Recorded by His Contemporaries*, trans. L. Sherley-Price (London: Mowbray, 1959), see especially p. 145.

20. Picola della Mirandola, *Oration on the Dignity of Man*.

21. Marsilio Ficino, *Theologica Platonica* III, 2 and XVI, 3; see also Giannozzo Manetti, *The Dignity and Excellence of Man*.

22. E. McCurdy, *The Mind of Leonardo da Vinci* (London: Cape, 1932), p. 78.

23. "Apology for Raimond de Sebonde," in his *Essays*.

24. *Discourse on Method*, vol. 5; see also his letter to Henry More, February 5, 1649. I have given the standard reading of Descartes, the way in which his position was understood at the time, and has been understood by most of his readers up to and including the present; but it has been claimed recently that this standard reading is a mistake, in that Descartes did not intend to deny that animals could suffer. For further details, see John Cottingham, "'A Brute to the Brutes?' Descartes' Treatment of Animals," *Philosophy* 53: 551–559 (1978).

25. John Passmore describes the question "why do animals suffer?" as "for centuries, the problem of problems. It engendered fantastically elaborate solutions. Malebranche [a contemporary of Descartes] is quite explicit that for purely theological reasons it is necessary to deny that animals can suffer, since all suffering is the result of Adam's sin and the animals do not descend from Adam." See John Passmore, *Man's Responsibility for Nature*, p. 114n.

26. Letter to Henry More, February 5, 1649.

27. Nicholas Fontaine, *Memories pour servir a l'histoire de Port-Royal* (Cologne, 1738), 2: 52–53; quoted in L. Rosenfield, *From*

Beast-Machine to Man-Machine: The Theme of Animal Soul in French Letters from Descartes to La Mettrie (New York: Oxford University Press, 1940).

28. *Dictionnaire Philosophique*, s.v. "Bêtes."

29. *Enquiry Concerning the Principles of Morals*, chapter 3.

30. *The Guardian*, May 21, 1713.

31. *Elements of the Philosophy of Newton*, vol. 5; see also *Essay on the Morals and Spirit of Nations*.

32. *Emile*, Everyman's Library (London: J. M. Dent & Sons), 1957, 2:118–120.

33. *Lecture on Ethics*, trans. L. Infield (New York: Harper Torchbooks, 1963), pp. 239–240.

34. *Hansard's Parliamentary History*, April 18, 1800.

35. E. S. Turner, *All Heaven in a Rage*, p. 127. Other details in this section come from chapters 9 and 10 of this book.

36. It has been claimed that the first legislation protecting animals from cruelty was enacted by the Massachusetts Bay Colony in 1641. Section 92 of "The Body of Liberties," printed in that year, reads: "No man shall exercise any Tirranny or Crueltie towards any bruite Creature which are usuallie kept for man's use"; and the following section requires a rest period for animals being driven. This is a remarkably advanced document; one could quibble over whether it was technically a "law," but certainly Nathaniel Ward, compiler of the "Body of Liberties," deserves to be remembered along with Richard Martin as a legislative pioneer. For a fuller account, see Emily Leavitt, *Animals and Their Legal Rights* (Washington, D. C.: Animal Welfare Institute, 1970).

37. Quoted in E. S. Turner, *All Heaven in a Rage*, p. 162. For an exploration of the implications of this remark that is a valuable sup-

plement to the discussion here, see James Rachels, *Created From Animals: The Moral Implications of Darwinism* (Oxford: Oxford University Press, 1990).

38. Charles Darwin, *The Descent of Man* (London, 1871), p. 1.

39. Charles Darwin, *The Descent of Man*, p. 193.

40. See Lewis Gompertz, *Moral Inquiries on the Situation of Man and of Brutes* (London, 1824); H. S. Salt, *Animals' Rights* (London, 1892; new edition, Clark's Summit Pennsylvania, Society for Animal Rights, 1980) and other works. I am indebted to *Animals' Rights* for some of the quotations in the following pages.

41. Book 2, chapter 11; for the same idea, see Francis Wayland, *Elements of Moral Science* (1835), reprint, J. L. Blau, ed. (Cambridge: Harvard University Press, 1963), p. 364, perhaps the most widely used work on moral philosophy in nineteenth-century America.

42. Quoted by S. Godlovitch, "Utilities," in Stanley and Roslind Godlovitch and John Harris, eds., *Animals, Men and Morals* (New York: Taplinger, 1972).

43. Quoted in H. S. Salt, *Animals' Rights*, p. 15.

44. Benjamin Franklin, *Autobiography* (New York: Modern Library, 1950), p. 41.

45. Quoted in H. S. Salt, *Animals' Rights*, p. 15.

46. *La Bible de l'humanité*, quoted in H. Williams, *The Ethics of Diet* (abridged ed., Manchester and London, 1907), p. 214.

47. *On the Basis of Morality*, trans. E. F. J. Payne (Library of Liberal Arts, 1965), p. 182; see also *Pargera und Paralipomena*, chapter 15.

48. See E. S. Turner, *All Heaven in a Rage*, p. 143.

49. E. S. Turner, *All Heaven in a Rage*, p. 205.

50. T. H. Huxley, *Man's Place in Nature* (Ann Arbor: University of Michigan Press, 1959), chapter 2.

Chapter 6

1. Dean Walley and Frieda Staake, *Farm Animals* (Kansas City: Hallmark Children's Editions, no date).

2. M. E. Gagg and C. F. Tunnicliffe, *The Farm* (Loughborough, England: Ladybird Books, 1958).

3. An example: Lawrence Kohlherg, a Harvard psychologist noted for his work on moral development, relates how his son, at the age of four, made his first moral commitment, and refused to eat meat because, as he said, "it's bad to kill animals." It took Kohlherg six months to talk his son out of this position, which Kohlherg says was based on a failure to make a proper distinction between justified and unjustified killing, and indicates that his son was only at the most primitive stage of moral development. (L. Kohlherg, "From Is to Ought," in T. Mischel, ed., *Cognitive Development and Epistemology*, New York: Academic Press, 1971, pp. 191–192.) Moral: If you reject a pervasive human prejudice, you just can't be morally developed.

4. W. L. Gay, *Methods of Animal Experimentation* (New York: Academic Press, 1965), p. 191; quoted in Richard Ryder, *Victims of Science* (London: Davis-Poynter, 1974).

5. Bernhard Grzimek, "Gequälte Tiere: Unglück für die Landwirtschaft," in *Das Tier* (Bern, Switzerland), special supplement.

6. Examples are the 1876 British Cruelty to Animals Act and the 1966–1970 Animal Welfare Act in the United States, both of which were enacted in response to concern about animals being used in experiments but have done little to benefit those animals.

7. For a list of some of the more radical organizations, see Appendix 3.

8. E. S. Turner, *All Heaven in a Rage* (London: Michael Joseph, 1964), p. 129.

9. E. S. Turner, *All Heaven in a Rage*, p. 83.

10. Gerald Carson, *Cornflake Crusade* (New York: Rinehart, 1957), pp. 19, 53–62.

11. E. S. Turner, *All Heaven in a Rage*, pp. 234–235; Gerald Carson, *Men, Beasts and Gods* (New York: Scribner's 1972), p. 103.

12. See Farley Mowat, *Never Cry Wolf* (Boston: Atlantic Monthly Press, 1963), and Konrad Lorenz, *King Solomon's Ring* (London: Methuen, 1964), pp. 186–189. I owe the first reference to Mary Midgley, "The Concept of Beastliness: Philosophy, Ethics and Animal Behavior," *Philosophy* 48: 114 (1973).

13. See, in addition to the references above, works by Niko Tinbergen, Jane van Lawick-Goodall, George Schaller, and Irenaus Eibl-Eibesfeldt.

14. See pages 207–208 above.

15. See page 208.

16. See Judy Mann, "Whales, Hype, Hypocrisy," *The Washington Post*, October 28, 1988.

17. I am often asked: What should we do about our cats and dogs? Some vegetarians are understandably reluctant to buy meat for their companion animals, for to do so is still to support the exploitation of animals. In fact it is not difficult to raise a dog as a vegetarian—Irish peasants who could not afford meat did it on milk and potatoes for centuries. Cats present more of a problem, since they need taurine, an amino acid not readily available from plants. It is now possible, however, to obtain a vegetarian taurine supplement from the American group, Harbinger of a

New Age. It has been claimed that this makes it possible for cats to be healthy on a vegetarian diet, but the health of cats on such a diet should be watched closely. Information can also be obtained from the British Vegetarian Society. See Appendixes 2 and 3 for addresses.

18. "On the Legality of Enslaving the Africans," by a Harvard student; quoted in Louis Ruchames, *Racial Thought in America* (Amherst: University of Massachusetts Press, 1969), pp. 154–156.

19. See Leslie Stephen, *Social Rights and Duties* (London, 1896) quoted by Henry Salt, "The Logic of the Larder," which appeared in Salt's *The Humanities of Diet* (Manchester: The Vegetarian Society, 1914), pp. 34–38, and has been reprinted in T. Regan and P. Singer, eds., *Animal Rights and Human Obligations* (Englewood Cliffs, N. J.: Prentice-Hall, 1976).

20. S. F. Sapontzis has argued that the possible happy life of a normal child and the possible miserable life of a deformed child are both reasons for having or not having the child only when the child is in existence, and there is thus no asymmetry. (S. F. Sapontzis, *Morals, Reason and Animals*, Philadelphia: Temple University Press, 1987, pp.193–194.) But this would mean that it is not wrong to decide to conceive a miserable child, although it is wrong to decide to keep the child alive once it exists. What if one knows, at the time the child is conceived, that one will have no opportunity for having an abortion or for carrying out euthanasia after the child is born? Then there will be a miserable child, so it would seem that a wrong has been done. But in Sapontzis's view, there appears to be no time at which that wrong can be done. I cannot see that this suggestion solves the problem.

21. See my *Practical Ethics* (Cambridge: Cambridge University Press, 1979) chapters 4 and 6. For further discussion, see Michael Lockwood, "Singer on Killing and the Preference for Life," *Inquiry* 22 (1–2): 157–170; Edward Johnson, "Life, Death and Animals," and Dale Jamieson, "Killing Persons and Other Beings," both in Harlan Miller and William Williams, eds., *Ethics and Animals* (Clifton, N. J.: Humana Press, 1983); Johnson's essay has been reprinted in T. Regan and P. Singer, eds., *Animal Rights and*

Human Obligations (Englewood Cliffs, N. J.: Prentice Hall, 2nd edition, 1989). See also S. F. Sapontzis, *Morals, Reason and Animals*, chapter 10. To understand the arguments lurking behind the entire debate, however, the indispensable (but not easy!) source is Derek Parfit, *Reasons and Persons* (Oxford: Clarendon Press, 1984), part IV.

22. The leading defender of rights for animals is Tom Regan; see his *The Case for Animal Rights* (Berkeley and Los Angeles: University of California Press, 1983). I have indicated why I differ in "Utilitarianism and Vegetarianism," *Philosophy and Public Affairs* 9: 325–337 (1980); "Ten Years of Animal Liberation," *The New York Review of Books*, April 25, 1985; and "Animal Liberation or Animal Rights," *The Monist* 70: 3–14 (1987). For a detailed argument that a being without a capacity to see itself as existing over time cannot have a right to life, see Michael Tooley, *Abortion and Infanticide* (Oxford: Clarendon Press, 1983).

23. A defense of such a position is presented in R.M. Hare's forthcoming article, "Why I Am Only a Demi-vegetarian."

24. Brigid Brophy, "In Pursuit of a Fantasy," in Stanley and Roslind Godlovitch and John Harris, eds., *Animals, Men and Morals* (New York: Taplinger, 1972), p. 132.

25. See Cleveland Amory, *Man Kind?* (New York: Harper and Row, 1974), p. 237.

26. Lewis Gompertz, *Moral Inquiries on the Situation of Man and of Brutes* (London, 1824).

27. For a powerful account of the cruelty inherent in the Australian wool industry, see Christine Townend, *Pulling the Wool* (Sydney: Hale and Iremonger, 1985).

28. See Appendix 2.

29. For examples of how brutal and painful the killing of "pests" can be, see Jo Olsen, *Slaughter the Animals, Poison the Earth*, (New York: Simon and Schuster, 1971) pp. 153–164.

30. A few scattered researchers have now begun work on contraception for wild animals; for a review, see J. F. Kirkpatrick and J. W. Turner, "Chemical Fertility Control and Wildlife Management," *Bioscience* 35: 485–491 (1985). But the resources being put into this area remain minuscule in comparison with those spent on poisoning, shooting, and trapping.

31. *Natural History* 83 (3): 18 (March 1974).

32. In A. I. Melden, ed., *Human Rights* (Belmont, Calif.: Wadsworth, 1970), p. 106.

33. Frankena, in *Social Justice*, p. 23.

34. H. A. Bedau, "Egalitarianism and the Idea of Equality," in J. R. Pennock and J. W. Chapman, eds., *Nomos IX: Equality* (New York, 1967).

35. G. Vlastos, "Justice and Equality," in *Social Justice*, p. 48.

36. J. Rawls, *A Theory of Justice* (Cambridge: Harvard University Press, Belknap Press, 1972), p. 510. For another example, see Bernard Williams, "The Idea of Equality," in P. Laslett and W. Runciman, eds., *Philosophy, Politics and Society*, second series, (Oxford: Blackwell, 1962), p. 118.

37. For an example, see Stanley Benn's "Egalitarianism and Equal Consideration of Interests," *Nomos IX: Equality*, pp. 62ff.

38. See Charles Magel, *Keyguide to Information Sources in Animal Rights* (Jefferson, N.C.: McFarland, 1989). The works of just a few of these philosophers are listed in Appendix 1.

39. R. G. Frey, "Vivisection, Morals and Medicine," *Journal of Medical Ethics* 9: 95–104 (1983). Frey's major critique of my work is *Rights, Killing and Suffering* (Oxford: Blackwell, 1983), but see also his *Interests and Rights: The Case Against Animals* (Oxford: Clarendon Press, 1980). I respond (too briefly) to these books in "Ten Years of Animal Liberation," *The New York Review of Books*, April 25, 1985.

40. See M. A. Fox, *The Case for Animal Experimentation* (Berkeley: University of California Press, 1986) and Fox's letter in *The Scientist,* December 15, 1986; see also Fox's "Animal Experimentation: A Philosopher's Changing Views," *Between the Species* 3: 55–60 (1987), and the interview with Fox in *Animals' Agenda,* March 1988.

41. Katherine Bishop, "From Shop to Lab to Farm, Animal Rights Battle is Felt," *The New York Times,* January 14, 1989.

42. "The Battle Over Animal Rights," *Newsweek,* December 26, 1988.

43. See Henry Spira, "Fighting to Win" in Peter Singer, ed., *In Defense of Animals* (Oxford: Blackwell, 1985), pp. 194–208.

44. See Alex Pacheco with Anna Francione, "The Silver Spring Monkeys," in Peter Singer, ed., *In Defense of Animals,* pp. 135–147.

45. See Chapter 2, note 118.

46. *Newsweek,* December 26, 1988, pp. 50–51.

47. Barnaby J. Feder, "Research Looks Away From Laboratory Animals," *The New York Times,* January 29, 1989, p. 24; for an earlier picture of the work of the Coalition to Abolish the LD50 and Draize Tests, see Henry Spira, "Fighting to Win" in Peter Singer, ed., *In Defense of Animals.*

48. Government of Victoria, *Prevention of Cruelty to Animals Regulations,* 1986, no. 24. The regulation covers the testing of any chemical, cosmetic, toilet, household, or industrial preparation. It prohibits the use of the conjunctival sac of rabbits for that purpose, and also prohibits any test in which animals are subjected to a range of increasing doses and the number of fatalities is used for the purpose of producing a statistically valid result. On New South Wales, see *Animal Liberation: The Magazine* (Melbourne) 27: 23 (January–March 1989).

Index

Peter Singer is Professor of Philosophy and Director of the Centre for Human Bioethics at Monash University, Melbourne. He is a frequent contributor to *The New York Review of Books* and to several other magazines and scholarly journals. He is the author of the major article on Ethics in the current edition of the *Encyclopaedia Britannica.* Professor Singer has taught at University College, Oxford, New York University, the University of Colorado at Boulder, and the University of California at Irvine. In addition to his academic activities, Professor Singer is actively involved in the Animal Liberation movement, as President of Animal Liberation (Victoria) and Vice-President of the Australian and New Zealand Federation of Animal Societies.